Ana Paula Bandeira de Oliveira

P + L Concepts for Food and Nutrition Units

AF306926

Ana Paula Bandeira de Oliveira

P + L Concepts for Food and Nutrition Units

Printed by Books on Demand GmbH, Norderstedt / Germany

Ana Paula Bandeira de Oliveira

P + L Concepts for Food and Nutrition Units

Application of P+L concepts in a UAN - Case study

Imprint

Any brand names and product names mentioned in this book are subject to trademark, brand or patent protection and are trademarks or registered trademarks of their respective holders. The use of brand names, product names, common names, trade names, product descriptions etc. even without a particular marking in this work is in no way to be construed to mean that such names may be regarded as unrestricted in respect of trademark and brand protection legislation and could thus be used by anyone.

Cover image: www.ingimage.com

This book is a translation from the original published under ISBN 978-613-9-66891-5.

Publisher:
Sciencia Scripts
is a trademark of
Dodo Books Indian Ocean Ltd. and OmniScriptum S.R.L publishing group

120 High Road, East Finchley, London, N2 9ED, United Kingdom
Str. Armeneasca 28/1, office 1, Chisinau MD-2012, Republic of Moldova, Europe
Printed at: see last page
ISBN: 978-620-8-10925-7

Copyright © Ana Paula Bandeira de Oliveira
Copyright © 2024 Dodo Books Indian Ocean Ltd. and OmniScriptum S.R.L publishing group

"It doesn't matter

Try again

Fail again

Failed better"

Samuel Beckett

ACKNOWLEDGEMENTS

To God, for life and for always guiding me along the best path.

To my family, especially my mother Eva, for her support, encouragement and sacrifices, believing in education as an instrument for building a better world.

To my beloved husband Fabrício, for his unconditional support, patience and words of encouragement in times of discouragement.

My greatest asset, my children Pedro and Mariana. A divine blessing in my life.

To Professor Dr Delmar Bizani (UNILASALLE), for his guidance, availability and attention at all times.

Dear Master Regina Alcântara (UNISINOS), for her inspiration and for believing in my potential since my undergraduate days, with countless words of affection, encouragement and motivation in building my academic and professional career.

To everyone in my life who encouraged me during the most difficult times.

To the professional nutritionists, from the most diverse UAN's, who directly or indirectly contributed to the construction of this work.

SUMMARY

The Food and Nutrition Unit is a group of areas with the aim of operationalising the nutritional provision of communities. In Brazil, the UAN segment feeds more than nine million people a day. Consuming more than three thousand tonnes of food per day and generating waste as a result of this production process. Minimising waste at source is one of the actions proposed by P+L and its approach determines techniques and technologies for sustainable development. The benefits of P+L come from actions that seek to reduce, minimise or eliminate raw materials and inputs that have a negative impact on the environment, generating more products and less waste, reducing environmental liabilities and improving health and safety at work. This work was carried out in a UAN serving 300 meals a day in the municipality of Porto Alegre - RS, with the aim of applying P+L concepts in a UAN, where solid waste from the production process of this menu was collected at 5 points during the months of October 2015 to March 2016. The following points were monitored: "unused/wasted", "pre-preparation" and "edible and inedible preparation", "clean leftovers", "burnt oil" and "coffee grounds". The data was organised, categorised and analysed in the light of P+L concepts. The treatment and analysis of the results of this work, in which data was collected over six months, generated twenty-one suggestions for improvement, four of which were implemented by the UAN, such as: a) stock control adjusted to production; b) acquisition of processed items for salads and garnishes; c) monitoring of oil saturation; and, d) menu planning based on the user profile. During the period, the collection points generated 1,075 kg to serve 34,736 meals. For the unused/old collection point, a 45 % reduction in waste generation was achieved. Comparing the month with the highest number of kilos found with the month with the lowest value, there was a 12% variation in monthly waste generation at the clean leftovers collection point. It is important to emphasise that this collection point directly reflects the financial loss of the meal produced and not invoiced. This makes an important contribution to the literature in this area, since the greatest diversity of work refers to the treatment of waste generated at the end of the process. In Brazil, the management of UANs has undergone changes in legislation and market behaviour, with increasing questions being raised about the control and inspection mechanisms for the waste generated in this production process and its environmental impacts. It is necessary to change the behaviour adopted. Even though it has technology, it is a segment that depends on the behaviour of its operational staff, and recognising this information, we see opportunities for training, process modification and the adoption of good practices.

Keywords: Food and nutrition units. Organic waste. Cleaner production.

SUMMARY

CHAPTER 1   5

CHAPTER 2   8

CHAPTER 3   9

CHAPTER 4   12

CHAPTER 5   30

CHAPTER 6   35

CHAPTER 7   49

CHAPTER 1

INTRODUCTION

In Brazil, according to data from the Brazilian Association of Collective Meals Companies (ABERC, 2009), the collective meals market as a whole provided 9.4 million meals/day.

Due to this demand, the sector is responsible for significant waste generation, mainly organic, effluent generation and water and electricity consumption. These impacts jeopardise the soil, water resources and atmosphere.

The activity has a significant economic turnover and its raw material is based on food of the most different categories. It consumes 3,000 tonnes of food every day and represents an annual revenue of one billion reais for governments in terms of taxes and contributions (ABERC, 2009).

According to Teixeira et al. (1990, p. 167),

> Food and nutrition units are organisations with a simple administrative structure but complex operations, since they generally carry out activities that fall within the technical, administrative, commercial, financial, accounting and security functions.

This approach is based on the understanding that evaluating the processes developed within it is fundamental to superior performance, "[...] finding an appropriate balance between economic growth and the preservation of natural resources, an objective known as sustainable development" (TOMAS; CALLAN, 2007, p. 24).

The initial tool in the production process of the Food and Nutrition Unit (FNU) is the menu, which "[...] serves as a management tool for the restaurant. From its planning, human and material resources can be dimensioned, costs can be controlled, purchases can be planned," (ABREU; SPINELLI, 2009) as well as the natural resources to be used and the generation of waste from this process. In this scenario, the contribution of,

> Cleaner Production (P+L) is a process of continuous improvement, which aims to make productive activity less damaging to the environment. The fundamental mechanism for making improvements is not only technology, but also a change in the management of the company and the people involved in the processes (SENAI/RS, 2003).

This approach proposes [...] so-called "end-of-pipe" management is extremely inefficient, as it causes waste and loss of raw materials in the process and then increases the cost of disposing of the waste generated as a result of the same production process. A change of focus is essential (MENDES, 2009).

As Wilkinson (apud MEDEIROS et al., 2010) summarises, source reduction is more than an economic incentive or a regulatory requirement. It is an environmental management priority that must be continuously measured.

P+L aims to identify, understand and answer these questions:

a) the nature of the waste and its emissions?

b) seek action rather than reaction;

c) preventing the generation of waste, effluents and emissions at source;

d) environmental problems are solved at all levels and with everyone's involvement.

Risks are reduced and transparency increased; its approach determines production techniques and technologies for sustainable development (SENAI/RS, 2003).

According to the National Centre for Clean Technologies (SENAI/RS, 2003), reducing waste and emissions also reduces waste management costs. Emphasise gains in productivity, indirectly collaborating with worker well-being.

Data shows that there is a significant loss in the collective meal production process. The figures show that establishments throw away around 15 per cent of what they prepare to serve users, equivalent to 5 per cent of gross sales. Others may lose 50 per cent of everything they produce, representing 15 per cent of their monthly turnover (ABRASEL apud BRADACZ, 2003, p. 13).

From the reception of raw materials to the distribution of ready-to-eat food, there are various types of control in a UAN. At each stage of the process, specific controls ensure greater practicality, generate workflow and come as close as possible to the ideal within the general conditions of task development (PROENÇA, 2005).

Comparing, analysing and evaluating the execution and performance of UAN services, the control of numerous tasks, i.e. quantity, quality, stock levels, deadlines, costs, product and service characteristics, hygiene, etc. (PROENÇA, 2005). Without this information, there is no control over production processes. These records are essential for implementing rationalisation measures, reducing waste and optimising productivity (ABREU; SPINELLI; ZANARDI, 2003).

The results of investments in work standardisation can be seen in the guarantee of quality services, gains in productivity, a reduction in waste and low-cost meals.

The management of operating units in Brazil has been strongly impacted by changes in legislation and market behaviour, where the waste generated at the end of the process has been questioned as to its destination, treatment and monitoring of the environmental impact (GRSA, 2013).

Today in a food and nutrition unit, the generation of organic waste is approximately 0.2kg/per capita. Studies report "[...] that the sum of waste of organic origin has reached 89 per cent." (SALES apud ALBERTONI, 2009).

According to the CNTL (2008), P+L means applying an integrated environmental, technical and economic strategy to processes and products in order to increase efficiency in the use of raw materials, water and energy, through the non-generation, minimisation or recycling of waste generated, with environmental and economic benefits for production processes.

The P + L concept considers the environmental variable at all levels of the organisation, developing products and processes, improving existing processes, aiming to reduce the negative environmental impacts generated as a whole, relating environmental issues to economic gains for the entire company (CNTL, 2008).

The benefits of this concept come from actions that seek to reduce and eliminate waste from processes, minimise or eliminate raw materials and inputs that have a negative impact on the environment, generating more products and less waste, reducing waste and emissions, energy efficiency, products and packaging, reducing environmental liabilities, and improving health and safety at work (CNTL, 2008).Since 1975, the Industries Centre for the Environment of the United Nations Environment Programme (UNEP) has been promoting actions for cleaner and safer production and consumption to achieve sustainable development.

To achieve this goal, UNEP has formed partnerships with industries, government sectors and international non-governmental organisations (UNEP/UNIDO, 2004).

In 1989, the concept of P+L was disseminated by UNEP as: P+L is the continuous application of an integrated and preventive environmental strategy, applied to processes, products and services, to increase eco-efficiency and reduce risks to man and the environment. It applies to:

- **production processes:** conserving raw materials and energy, eliminating toxic raw materials and reducing the quantity and toxicity of all emissions and waste;
- **products:** reducing the negative impact throughout the product's life cycle, from raw material extraction to final disposal;
- **services:** incorporating environmental concepts into the design and distribution of services. (UNITED, 2004).

P+L emphasises a change in the way of thinking about environmental issues and induces companies to find solutions that replace conventional "end of pipe" treatments with optimisation in production processes. These optimisations can occur through different forms of intervention in the production process, which include managerial and technological aspects and can range from improving operating and maintenance procedures (good operating practices) to changes in processes and products and technological innovations (MARINHO; KIPERSTOK, 2001).

When a P+L programme is implemented, cost savings are expected as a result of reduced consumption of raw materials, water and energy and minimising the generation of waste at source. In other words, there is an increase in the productivity of resources, generating environmental benefits and, consequently, economic advantages (NASCIMENTO; LEMOS; MELLO, 2002).

In this context, P+L has established itself as a useful tool, as it increases the efficiency of the production process, improves the competitiveness of organisations and rationalises the consumption of natural resources (CNTL, 2008).

CHAPTER 2

OBJECTIVES

2.1 General

The overall aim of the research was to apply P+L concepts in a Food and Nutrition Unit in order to reduce the environmental impact of the resources used and the waste generated during the manufacturing process of the final meal.

22 **Specific**

a) Characterise the Food and Nutrition Unit according to its production process;

b) Apply concepts from level 1 of P + L

CHAPTER 3

BACKGROUND

In Brazil, the Household Budget Survey (POF) indicated that in 2002-2003 the population spent 24 per cent of total expenditure on food on consumption outside the home. In the 2008-2009 survey, this expenditure exceeded 31 per cent (INSTITUTO BRASILEIRO DE GEOGRAFIA E ESTATÍSTICA, 2010).

As shown in figure 1, in the southern region of Brazil, the amount spent on eating out rose from 23.3 per cent in POF 2002-2003 to 27.7 per cent in POF 2008-2009.

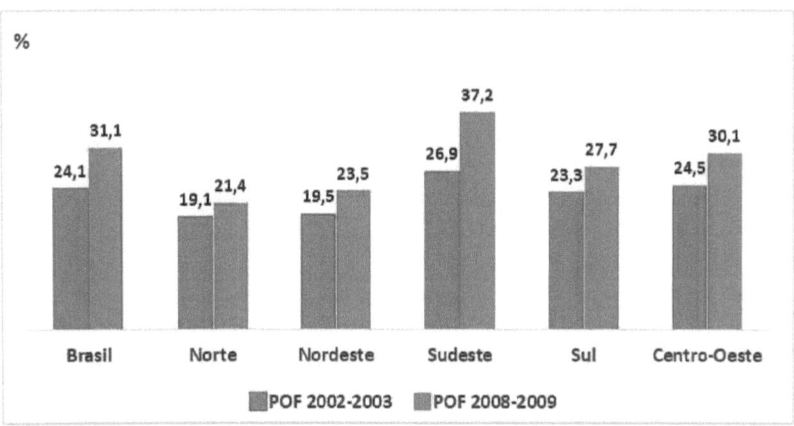

Figure 1 - POF - Period 2002/2009. Percentage of average monthly household monetary and non-monetary expenditure on food away from home, by major region

Source: IBGE, 2010.

A large part of these meals eaten away from home are produced in collective UANs, and these places should include in their objectives mechanisms aimed at reducing environmental impacts, as well as conserving natural resources. This segment contributes to the economic and social sphere by generating jobs and income for the economy (ABREU; SPINELLI, 2009; LEAL, 2010).

The importance of studying waste, not only in restaurants, is due to the fact that each inhabitant generates around 0.8kg of waste per day. Waste and soil contamination from improperly disposed of waste can be avoided by controlling the generation of this waste. In addition, failure to treat this waste has a significant impact on the health of the environment.

degradation of the biosphere and negatively affects the quality of life on the planet (SOUZA et al apud CARVALHO, 2011).

Various studies have been carried out in different regions of the country, as listed in Table 1, showing concern about the environmental aspects and impacts of UAN, but they refer to the "end of the pipe", where the waste generated is looked at.

Table 1 - Studies related to waste generation in UAN's in Brazil

References	States	Topics covered	Main features
Colares et al., 2008	RJ	Solid waste generation identifies how waste is segregated	Reports that organic waste per meal is 0.28kg
Alves and Ueno, 2015	SP	Quantify the waste generated	Reduce waste using the 3R's principles
Sanchez, 2009	GO	Waste generation diagnosis and composting feasibility analysis	Collection from all areas involved Reports 42.9kg/day, 36% of which is in the pre-preparation area, 17% in the kitchen and 47% on trays
Chieregatto and Claro, 2008	SP	Investigates commercial restaurants, proposes reverse logistics	Owners claim lack of partners and defined logistics
Bilck et al., 2009	PR	Evaluate the by-products generated, their final disposal and look for possible alternatives to the problems encountered.	No percentage identification of waste
Carmo et al., 2009	MG	Composting restaurant waste for afforestation	Suggests composting for afforestation
Souza, 2008	DF	Evaluates the potential for generating solid waste from the university restaurant	81% of waste is organic Waste is separated
Chamberlem, Kinazs, and Campos, 2012	MT	Evaluates leftover intake and discarded leftovers in two units of transported meals	Measure meal produced x meal distributed

Source: authorship, 2016.

It is also justified that this study could encourage interest in environmental issues in the sector, with the aim of raising awareness of environmental impacts. As table 1 shows, we did not find a large number of studies on this subject in the literature.

Table 1 - Compilation of the papers found in the research databases

Database	Key words	Frequency of work
ELSEVIER JOURNALS	*Cleaner production*	*288*
	Food and nutrition unit	*247*
	Minimisation of organic residues infood and nutrition unit	*5*
	Cleaner production in food and nutrition units	*3*
	Identification ofsolid organic waste	*200*
	Environmental impact in UAN	3
	Environmental impact in restaurants	4
	Minimising organic solid waste in UAN	Unregistered
JOURNAL DATABASE CAPES	Cleaner production	183
	Cleaner production in UAN	1
	Cleaner production in restaurants	14

Identifying waste in restaurants	17
Identification of waste in UAN	2
Collective catering	349

Source: authorship, 2016.

Clean technologies are relevant and interdisciplinary subjects, as the concepts of P+L in industrial processes have brought important elements for minimising the waste generated (SENAI, 2003).

This topic is part of the Environmental Management and Technologies research programme of the Academic Master's Degree in Environmental Impact Assessment.

CHAPTER 4

THEORETICAL FRAMEWORK

For a better understanding of the aspects to be worked on in this project, this chapter will present the topics according to their relevance to the food sector.

4.1 UAN - Food and nutrition unit

The Food and Nutrition Unit (UAN) is a set of areas with the aim of operationalising the nutritional provision of communities. It consists of an organised service, comprising a sequence and succession of acts designed to provide balanced meals within dietary and hygienic standards, thus aiming to meet the nutritional needs of its clients in a way that fits within the financial limits of the institution (ABREU; SPINELLI; ZANARDI, 2003).

The Brazilian Association of Collective Meals Companies (ABERC, 2013) publishes an annual report on the number of self-managed meals (managed by the company itself), collective meals (service providers) and convention meals (tickets/coupons for commercial restaurants) served in Brazil. In 2010 there were 14.89 million meals served per day and in 2011 there were 16.65 million. In 2012 there were 17.41 million meals. It is estimated that the sector will continue to grow and, likewise, there will be an increase in the generation of solid waste.

According to Sales (2009), studies and research lead to the conclusion that Brazilians eat less food than they throw away. It is estimated that between 30 and 40 per cent of what is produced in Brazil goes to waste. There are several factors that determine such levels of waste, such as inadequate planning of the number of meals to be produced, variations in the number of customers each day and their food preferences, inadequate food production and portioning due to a lack of staff training.

According to Proença (2005), meals can be eaten away from home in two categories of Meal Producing Unit (MPU), which, depending on their relationship with diners, can be classified as commercial (relative turnover) or collective (total turnover), the latter also being referred to as UANs, which can be located in industries, hospitals, schools, religious communities and the armed forces, among other institutions.

Its activities consume raw materials, water, energy and effluents, for example, resources that need to be managed and conducted efficiently. It is understood that within the UAN process there are opportunities to improve environmental management across the board. As an example, the organic waste generated in the production of collective meals has historically been treated as an item for disposal. This does not provide a final solution to the problem of waste [...] in this case organic waste, which ends up increasing costs for the producer. On the other hand, the use of inadequate treatment results in a waste of capital and little protection for the environment. Waste treatment does not add value to the product and should therefore be considered a last resort. The most sensible option is to minimise this type of treatment, interfering in the system to avoid or reduce the amount of waste to be disposed of (GIANNETTI; ALMEIDA, 2006).

Antonius (apud TINOCO; KRAEMER, 1999, p. 89) reveals that, in general, environmental management can be conceptualised as the integration of organisational systems and programmes that enable: "[...] the development and use of appropriate technologies to minimise or eliminate industrial waste".

According to Tinoco and Kraemer (2011, p. 99), "the introduction of environmental practices can, on the other hand, lead to cost savings by improving process efficiency".

The author SENAI/RS,2003, in figure 2, explains that "the main difference between cleaner technologies and end-of-pipe control methods is temporal: cleaner technologies are preventive, applied to avoid future problems [...]".

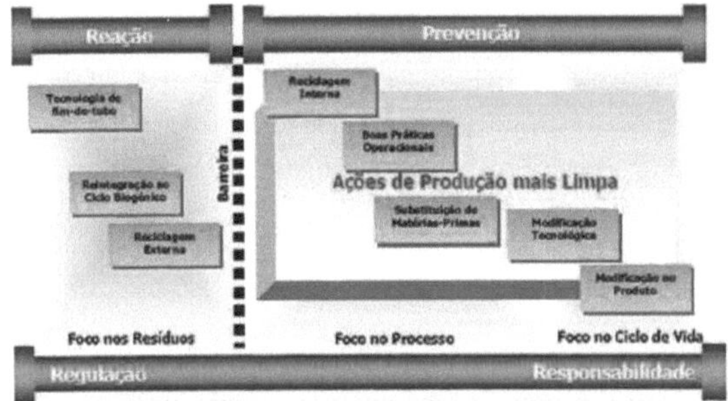

Figure 2 - Progress towards cleaner production (P + L)

Source: SENAI/RS, 2003.

42 Nutritionist work in food and nutrition units

As described in CFN (2005), in Art. I° , the nutritionist is a health professional who, in accordance with the principles of the science of nutrition, has the function of contributing to the health of individuals and the community.

The profession of nutritionist was created by Law No. 5.276 of 24 April 1967. On 17 September 1991, Law No. 8.234 regulated the profession of nutritionist and defined the professional's activities, which are: directing, coordinating and supervising undergraduate courses in nutrition; planning, organising, directing, supervising and evaluating food and nutrition services; planning, coordinating, supervising and evaluating dietetic studies; teaching professional subjects in undergraduate courses in nutrition; teaching nutrition and food subjects in undergraduate courses in the health area and other related areas; auditing, consultancy and advisory services in nutrition and dietetics; assistance and nutritional education to groups or individuals, healthy or sick, in public and private institutions and in nutrition and dietetics offices; dietotherapy assistance in hospitals, outpatient clinics and in nutrition and dietetics offices, prescribing, planning, analysing, supervising and evaluating diets for the sick (ASBRAN, 1991).

According to the CFN, Resolution No. 380 of 2005, these are the duties of the nutritionist by area of activity and legal basis:

Items II, VI and VII of Article 3° ; Items III, IV, XI and Sole Paragraph of Article 4° of Law No. 8.234/91. 1. FOOD AND NUTRITION UNIT (FNU) - It is the responsibility of the Nutritionist to plan, organise, direct, supervise and evaluate food and nutrition services when carrying out their duties in Food and Nutrition Units. Carry out assistance and nutritional education to healthy or sick individuals or groups in public and private institutions. [...Jl.1.5 Planning, coordinating and supervising the activities of selecting suppliers, the origin of food, as well as the purchase, receipt and storage of food; [...]; 1.1.7 Planning, implementing, coordinating and supervising the activities of pre-preparation, preparation, distribution and transport of meals and/or culinary preparations; [,...]; 1.1.8....]1.1.20. Carry out periodic checks on the work carried out; [...]Implement and supervise periodic checks on leftovers, waste-ingestion and waste analysis, promoting social, ecological and environmental awareness. (CFN, 2005).

In this study [...] the nutritionist's area of expertise is Collective Nutrition. According to Interministerial Ordinance No. 66 of 25 August 2006, which alters the nutritional parameters of the Workers' Food Programme (PAT), the person in charge of the PAT is a professional legally qualified in Nutrition, whose commitment is to correctly carry out the programme's nutritional activities, with a view to promoting healthy food for workers (CARVALHO, 2011 p. 28).

43 History of the Collective Meals sector

In 1976, the PAT was established by Law No. 6.321 of 14/04/1976 and regulated by Decree No. 5 of 14/01/1991, prioritising low-income workers. This programme is structured as a partnership between the government, the company and the worker, and its aim is to improve the nutritional conditions of workers, with positive repercussions on quality of life, a reduction in accidents at work and an increase in productivity; all of these benefits are not just for the worker, but also for the company and the federal government. In the 1980s, the Brazilian Association of Collective Meal Companies (ABERC), which brings together some of the companies responsible for providing collective meals in Brazil, was founded in 1984, and the Brazilian Association of Bars and Restaurants (ABRASEL) in January 1986 (CORRÊA; LANGE, 2011).

According to data from ABERC (2015), the size and importance of the sector in the national economy can be measured from the figures generated by the segment in 2009. It generates 9.8 billion reais a year, provides 180,000 direct jobs, consumes 3,000 tonnes of food a day and represents an annual revenue of one billion reais for governments in terms of taxes and contributions. Of the 9.4 million meals in 2009, approximately seven million were provided by the 90 service providers affiliated to ABERC, which together account for 93 per cent of the volume of this market (ABERC, 2015).

It is estimated that the theoretical potential for collective meals in Brazil is over 41 million units per day, which shows that the segment still has a lot to grow. The sector has managed to remain stable in recent years thanks in part to the process of outsourcing and the development of new market niches (CORRÊA; LANGE, 2011).

ABRASEL (2015), the representative of a sector that today brings together around one million companies and generates six million direct jobs throughout the country, has sought since its creation in 1986 to make an effective contribution to important advances in favour of the development of the out-of-home food segment on the national stage. This sector accounts for 2.4 per cent of the gross domestic product. In addition, the habit of eating away from home is growing and accounts for more than 25 per cent of Brazilians' spending on food (ABRASEL, 2015).

4.4 Legal requirements applicable to Food and Nutrition Units

The legal requirements followed by the segment are based on the Federal Constitution, the main legislation that plays a guiding role in protecting the environment. Article 225 states that those responsible for guaranteeing this protection are the state and society (BRASIL, 1988).

Law No. 6938/81 was the first to establish definitions, principles and targets for the preservation of the environment, and the National Environmental Council (CONAMA) is responsible for issuing resolutions on the licensing of polluting activities; another important piece of current Brazilian legislation on environmental sustainability is Law No. 12305/10, which instituted the National Solid Waste Policy (PNRS). The aim of the law is to put an end to rubbish dumps and collaborate with the inclusion of workers who collect recycled materials, making it possible to improve the social and economic conditions of these workers (MARTINS apud BRASIL, 2010).

Law 12.305/10, which establishes the PNRS, is very up-to-date and contains important instruments to enable the country to make the necessary progress in tackling the important environmental, social and economic issues arising from the inadequate management of solid waste (BRASIL, 2010).It provides for the prevention and reduction of waste generation, proposing the practice of sustainable consumption habits and a set of instruments to encourage increased recycling and reuse of solid waste (that which has economic value and can be recycled or reused) and the proper disposal of waste (that which cannot be recycled or reused). This law regulates the shared responsibility of waste generators: manufacturers, importers, distributors, traders, citizens and holders of urban solid waste management services in the reverse logistics of post-consumer waste and packaging (BRASIL, 2010).

Important targets have been set that will contribute to the elimination of rubbish dumps and establish planning instruments at all levels, as well as requiring private individuals to draw up their Solid Waste Management Plans (BRASIL 2010).

Martins (2015) pointed out that in the context of meal production, environmental sustainability can be described as ecologically sustainable practices that aim to mitigate environmental impact through the rational use of natural resources, reduce waste generation, increase recycling, encourage the use of agroecological foods, certify companies and carry out traceability of raw materials, as well as train employees, make use of more environmentally appropriate technologies and improve the implementation of environmental protection policies.

It can be seen that few collective restaurants collect and properly dispose of organic and recyclable waste, as well as implementing programmes to control water and energy consumption. Studies show that it is important to know the volume of recyclable materials, food waste and natural resources generated, in order to correctly dispose of waste and conserve resources such as water and energy (MARTINS, 2015).

CONAMA resolution no. 237 of 19/12/97 tells us in its article:

> Art. lo For the purposes of this Resolution, the following definitions are adopted: I - Environmental Licensing: administrative procedure by which the competent environmental body licenses the location, installation, expansion and operation of undertakings and activities that use environmental resources, considered to be effectively or potentially polluting or those that, in any form, may cause environmental degradation, considering the legal and regulatory provisions and technical standards applicable to the case. (BRASIL, 1997).

The activity of producing meals is not considered potentially polluting, so the UAN does not have this document (GRSA, 2013).

According to ABNT - NBR 10004 (2004) it is defined:

> 3.1 solid waste: Solid and semi-solid waste resulting from industrial, domestic, hospital, commercial, agricultural, service and sweeping activities. This definition includes sludge from water treatment systems, sludge generated in pollution control equipment and installations, as well as certain liquids whose particularities make it unfeasible to discharge them into the public sewage system or bodies of water, or require solutions that are technically and economically unfeasible in view of the best available technology.

Determined by CONAMA Resolution 01 of 08/03/1990 (BRASIL, 1990) and norm ABNT 10151 (ABNT, 2000), described:

> VI - For the purposes of this Resolution, measurements must be carried out in accordance with ABNT NBR 10.151 - Noise Assessment in Inhabited Areas for Community Comfort.
>
> Objectives of ABNT 101151
> 1.1 This Standard establishes the conditions required for assessing the acceptability of noise in communities, regardless of the existence of complaints. 1.2 This Standard specifies a method for measuring noise, the application of corrections to the measured levels if the noise has special characteristics and a comparison of the corrected levels with a criterion that takes several factors into account.

With regard to the PGRS, the UAN mentioned in this case study has a standard operating procedure, determined by health legislation RDC 216.

CONAMA 275 (BRAZIL, 2001) considers reducing the growing environmental impact associated with the extraction, generation, processing, transport, treatment and final disposal of raw materials, leading to an increase in rubbish dumps and landfills, to be a pressing need. Considering that environmental education campaigns, equipped with an easy-to-visualise identification system that is valid nationwide and inspired by forms of coding already adopted internationally, are essential to making selective waste collection effective, making it possible to recycle materials, it resolves in its articles cited below:

> Art.lo To establish the colour code for the different types of waste, to be adopted in the identification of collectors and transporters, as well as in information campaigns for selective collection. Art. 2 Selective collection programmes, created and maintained within the scope of federal, state and municipal public administration bodies, both direct and indirect, and parastatal entities, must follow the colour standard set out in the annex. § Paragraph 1 - The adoption of this colour code is recommended for selective collection programmes set up by the private sector, cooperatives, schools, churches, non-governmental organisations and other interested entities. § The organisations listed in the heading of this article will have up to twelve months to adapt to the terms of this Resolution. Art. 3 The inscriptions with the names of the waste and additional instructions regarding segregation or the type of material will not be standardised, but it is recommended that the colours black or white be adopted, according to the need for contrast with the base colour. Art. 4 This Resolution comes into force on the date of its publication (BRASIL, 2001).

CONAMA Resolution No. 430, issued on 13 May 2011, publishes:

> Art. lo This Resolution provides for conditions, parameters, standards and guidelines for managing the discharge of effluents into receiving bodies of water, partially amending and complementing Resolution 357 of 17 March 2005 of the National Environment Council (CONAMA).
> Sole Paragraph. The indirect discharge of effluents into the receiving body must comply with the provisions of this Resolution when it is verified that there is no specific legislation or standards, provisions of the competent environmental body, as well as guidelines from the operator of the sanitary sewage collection and treatment systems.
> Art. 2 The disposal of effluents on the ground, even if treated, is not subject to the parameters and discharge standards set out in this Resolution, but may not cause pollution or contamination of surface and underground waters. Art. 3 Effluents from any polluting source may only be discharged directly into receiving bodies after due treatment and provided that they comply with the conditions, standards and requirements set out in this Resolution and other applicable norms. (BRASIL, 2011).

Referring to the municipal level, in Porto Alegre, complementary law no. 728, from January 2014, establishes the municipal urban cleaning code, revokes complementary laws no. 234, of 10 October 1990, 274 of 25 March 1992, 376 of 3 June 1996, 377 of 3 June 1996, 591 of 23 April 2008 and 602 of 24 November 2008, and makes other provisions. Where in Section X, in the general rules it describes the following:

> Article 58: It is forbidden in the entire territory of the Municipality of Porto Alegre [...].
> Article 59: The use of *in natura* waste as feed for pigs or other animals is prohibited
> § If an irregularity is found, it must be reported to the competent public health bodies, so that the appropriate measures can be taken, without prejudice to the imposition of a fine (PORTO ALEGRE, 2014).

4.5 Cleaner production (P + L)

4.5.1 *History of cleaner production*

In 1994, UNIDO and UNEP created the P+L programme for environmental preservation. This programme is an integrated and preventive strategy that aims to increase company productivity, reduce the costs of raw materials, energy and natural resources and, consequently, reduce environmental impact in a sustainable way. In order to implement the programme and promote its application in companies and developing countries, there are around 31 National Programmes and National Cleaner Production Centres, located on various continents, whose main role is to promote demonstrations in the industrial plant; carry out training for all those involved; disseminate information specific to the programme and promote the evaluation of environmental policies (CNTL, 2008).

Figure 3 describes the countries where the P+L centres are located. In July 1995, the Brazilian NCPC was inaugurated, called the National Clean Technology Centre - CNTL SENAI, which is located at the National Industrial Learning Service - SENAI, in Porto Alegre, in the state of Rio Grande do Sul. The CNTL SENAI has the function of acting as a facilitator for the dissemination and implementation of the Cleaner Production centre in all productive sectors.

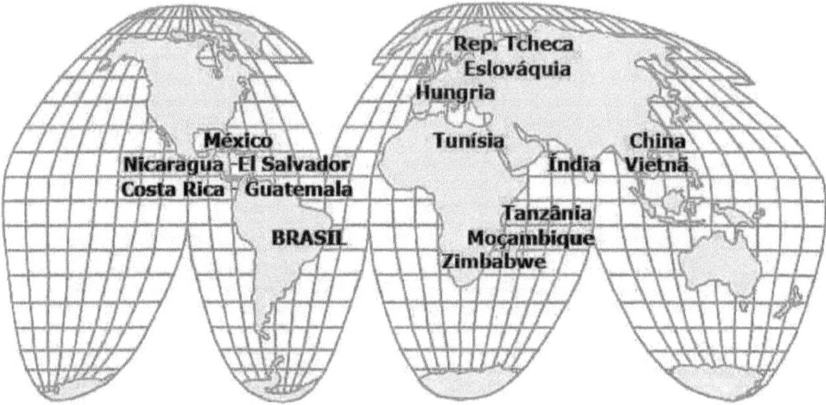

Figure 3 - R&D centres around the world
Source: CNTL, 2008.

The programme developed in Brazil is an adaptation of the UNIDO/UNEP programme and the experience of the Stenum Consultancy in the city of Graz, Austria, which developed the Ecological Project Integrated Environmental Technologies - ECOPROFIT.

4.5.2 Why invest in cleaner production?

Referring to known environmental problems, the Cleaner Production Programme investigates the production process and other activities of a company and studies them from the point of view of the use of materials and energy.

The P+L Programme aims to strengthen the industry economically through pollution prevention, embedded in the dream of contributing to environmental improvement in the region.

Supported by the introduction of innovations within the companies themselves, with the aim of leading

18

them towards sustainable development, where products, technologies and materials are carefully studied in order to minimise waste, emissions and effluents, and find ways to reuse unavoidable waste. In this sense, this programme is not a solution to an isolated problem, but a profitable tool for establishing a holistic concept (CNTL, 2008). Some of the reasons for implementing the programme are:

-lower   costs for production, end-of-pipe treatment, health care and total cleaning (removal of gases) of the environment;

-improved process efficiency and product quality, thus contributing to industrial innovation and competitiveness;

-reduction of   risks to workers, the community, consumers of products and future generations, thus reducing their costs of risks and insurance premiums;

-promote   the guarantee of the company's public image, producing intangible social and economic benefits.

According to the National Clean Technology Centre, two different but interrelated groups of P+L promoters can be identified. Firstly, there are the companies interested in P+L because the owners and employees are concerned about maintaining a clean, properly organised and environmentally friendly work area.

Secondly, there are companies where the adoption of P+L practices will be motivated by a reduction in operating costs, either by a reduction in waste or by a reduction in associated taxes. In addition, P+L gives companies a competitive advantage in markets where there is demand for environmentally improved products (SENAI, 2003).

Figure 4 shows how the P+L approach differs from the traditional (end-of-pipe) approach. P+L always focuses on the preventive side, seeking to avoid the problem, while the traditional approach focuses on solving the problem that has already been created.

Figure 4 - Differential approach of P+L and End-of-Pipe Techniques
Source: FIERGS; SENAI, 2008.

There will certainly be greater difficulties in answering the questions posed in P+L at first. However, when the questions are fully answered, a definitive solution will be suggested, which is the main objective. If the assessment is carried out with a focus on waste costs, the P+L solution will always be more economical in the long term, not least because it will be definitive and preventative; in other words, the waste will not be generated and therefore will not be handled, transported, stored or disposed of.

Consequently, there will be a reduction in the costs associated with waste. What's more, when processes become more efficient and raw materials are actually transformed into products, the amount of raw materials the company has to buy will also be reduced, as they will only be used to produce products and not the sum of products + waste.

At first, waste may not be completely eliminated, but only reduced; even so, economic and environmental benefits can be realised.

It is precisely the difference in approach between end-of-pipe technologies and P+L that makes it possible to obtain economic and environmental benefits for companies. The degree of complexity of the solutions is greater in P+L, as it penetrates deep into the company, into the way it carries out its activities, and requires massive support from employees. However, once this cultural change in the way problems are solved has been adopted and the search for continuous improvement has begun, everything else follows (CNTL, 2008).

According to DONADON (2005), the adoption of P+L in a juice and soft drink producing industry has enabled processes to be reviewed and benefits to be realised, such as disposing of the pomace and fruit peels left over from the product's manufacturing process for external recycling and earning an income from this input. Where before it was disposed of in landfill.

Another P+L action in this case was to change the fruit-picking process, which consumed electricity. A greater number of people carried out this task, increasing energy efficiency (where the same number of machines are switched on, with a greater quantity picked), obtaining better production performance

(DONADON, 2005).

Examples of economic and environmental benefits can be found in CNTL 2005, which reports that two bakeries in the municipality of Pelotas /RS achieved benefits after implementing reduction actions:

Bakery 1:

- reduction in the generation of flour waste: 224 g/year;
- reduction in flour consumption: 448 kg/year;
- reduction in firewood consumption by using pallet waste: 45 m3.

With economic results totalling over R$ 2,000.00 and with an immediate payback period.

Bakery 2:

- reduction in flour consumption by 2,300 kg/year;
- reduction in the generation of flour waste by 368 kg/year;
- reduction in electricity consumption by changing the temperature settings of refrigeration equipment, with a payback period of 5.2 years.

In the UAN segment, the development of waste management, with a Cleaner Production approach, ultimately leads to better organisation of the process, with a cleaner environment, tidier waste and a consequent reduction in accidents at work. In addition, the process makes it possible to quantify wasted materials and visualise the responsibilities for improving the production process (SENAI 2003,SENAI 2008).

According to figure 5, [...] in order to take P+L actions, the causes of waste generation and opportunities for modification are observed at various levels of strategy application (SILVA, 2010, p. 49).

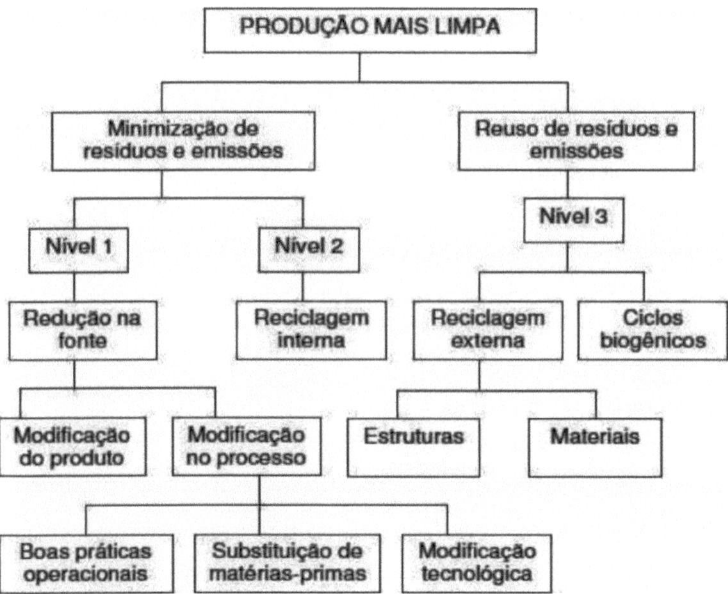

Figure 5 - P+L Application Levels
Source: CNTL apud RIEHL, 2008, p. 38.

4.6 Linear production flowchart

Determining the flow of raw materials, personnel and the use of equipment must take into account their intersections and interferences in terms of time, method, health and possible risks of food contamination (ABREU, 2009).

4.6.1 Receiving inspection

Management practices in receiving, such as checking purchase orders against what has been received, packaging integrity, shelf life, product characteristics, hygiene of the supplier and their vehicle, help to minimise possible future waste (ABREU; SPINELLI, 2007).As described in the GRSA Good Practices Manual (2015), the receiving inspection must be carried out when the product arrives. The restaurant must keep track of the products delivered by suppliers and record the checks.

The person responsible for this check is the nutritionist or another person designated by him/her.

4.62 Storage

Its aim is to establish criteria for the storage of raw materials, semi-finished products, ready-to-eat products, disposable materials and cleaning materials, in order to guarantee the original characteristics of the products, preventing microbiological, physical and organoleptic alterations (GRSA, 2015).

The causes of wastage at this stage include a lack of organisation of spaces (shelves, monoblocks); products without identification or with incomplete identification, leading to doubts about the expiry date; lack of cleaning; expired products; the presence of insects and rodents and insufficient maintenance of equipment temperatures (BRADACZ, 2003).

According to GRSA (2015), raw materials, disposables and cleaning products must be stored on pallets, boards and/or shelves, respecting the minimum spacing necessary to ensure adequate ventilation, cleaning and, where appropriate, disinfection of the area. Pallets, pallet racks and/or shelves must be made of smooth, resistant, waterproof and washable material. Shelves must not be made of wood.

Keeping the place organised and clean allows for a better view of possible pest and rodent infestations and makes it easier to keep the products in order on the shelves and racks, according to the PVPS (first in, first out) rules, which indicate that the food that is due to expire first should be placed in front of the rest so that it can be used first (BRADACZ, 2003).

All products must be stored at a minimum distance of 25 cm from the floor, 10 cm from the wall and between the piles of food and 60 cm from the ceiling and must never be stored directly on the floor, leaning against the wall or ceiling (GRSA, 2015).

The refrigeration equipment must be in line with the needs and types of food to be produced/stored.

Different types of food can be stored in the same unit, as long as they are properly packaged and separated. If there is only one piece of positive refrigeration equipment, set it at a maximum of +4° C, respecting the following layout:

-ready-to-eat food   on the top shelves;

-semi-ready and/or pre-prepared foods   on the middle shelves;

-fresh   food such as chilled cams, defrosted products, which will still be pre-prepared for consumption (GRSA, 2015).

According to the Good Practices Manual:

> Only open the doors of refrigeration equipment when necessary and for as short a time as possible, in order to optimise the equipment's capacity.
> Disposables should be stored separately from food products and sanitising and cleaning products.
> Cleaning and sanitising products should be stored separately from food products and disposables
> Before storage, the rooms must be thoroughly sanitised.
> Cam/jam storage: raw materials, ingredients and packaging inspected on receipt, following pre-established criteria for each product. (GRSA, 2015 p. 11).

Product labelling in accordance with specific legislation (RIO GRANDE DO SUL, 2009).

It must be stored in accordance with Ordinance 78/RS:

8.4. Temperature of raw materials, ingredients and industrialised products stored according to the manufacturer's instructions or according to the following criteria:
I. Frozen food: -18° C or lower;
II. Refrigerated food: below 5° C;
III. Existence of records proving temperature control in storage, checked, dated and initialled. Frozen food stored exclusively under freezing, chilled food stored exclusively under refrigeration, or according to labelling. (RIO GRANDE DO SUL, 2009, p. 38).

Sufficient refrigeration and freezing equipment for the needs and types of food to be stored (RIO GRANDE DO SUL, 2009).

The aim of exercising control over the various stages of the production process is to commit to a safe, quality meal. Figure 6 shows the stages of meal production (GRSA, 2015).

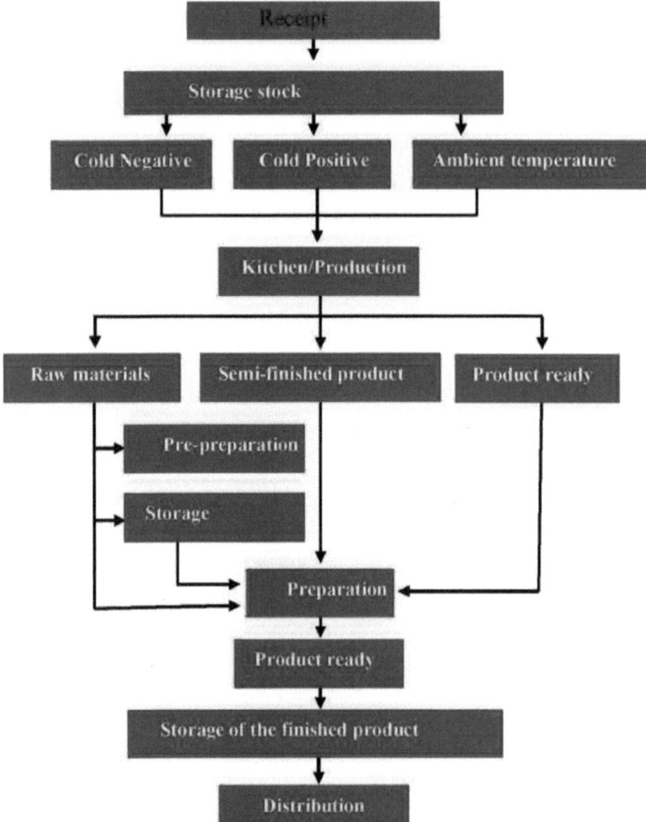

Figure 6 - Linear production flowchart
Source: GRSA, 2015.

4.6.3 Production process

For GRSA (2015), the description of activities is as follows:

1) Pre-preparation /Preparation
   a. Hygienisation of fruit and vegetables
   b. Defrosting
2) Cooking
3) Cooling

Control of the production process must be carried out at all stages of meal preparation. These stages must be controlled and monitored.

- <u>Pre-preparation/Preparation</u>

Pre-preparation - the stage where the food is modified by sanitising, seasoning, cutting, portioning, selecting, choosing, grinding and/or adding other ingredients (GRSA, 2015).

- <u>Sanitising fruit and vegetables</u>

The process of sanitising food is of fundamental importance in eliminating or reducing microbiological hazards, minimising the risk of transmitting pathogens that cause disease.

The "in natura" food arrives at the site with impurities and dirt. These particles carry bacteria inside which, if not properly eliminated through the washing and sanitising stages, can contaminate the food itself, the environment and other foods through cross-contamination.

According to GRSA (2015), this process involves selection, washing and disinfection.

Accordingly, Ordinance 78 (RIO GRANDE DO SUL, 2009, p. 39) describes:

> 9.22. The hygiene procedures for fruit and vegetable products follow the following criteria:
> I. Sorting food, removing parts or products that have deteriorated and are not in a suitable condition;
> II. Careful washing of food one by one with drinking water;
> III. Disinfection: immersion in a chlorine solution containing 100 to 250ppm of free chlorine for 15 minutes, or other suitable products registered with the Ministry of Health, authorised for this purpose and in accordance with the manufacturer's instructions;
> IV. Rinse with drinking water.

<u>   -Rinse</u>

Wash in running water (rinse) to remove chlorine residue, using a colander or sieve.

Product ready for preparation or direct consumption.

Once the above procedures have been carried out, the raw material is ready to be prepared according to the menu.

-Defrosting

The stage where the food goes from freezing temperature to up to 4°C under refrigeration (RIO GRANDE DO SUL, 2009).

Defrosting is carried out under refrigeration at a temperature below 5°C.

-Cocção

Stage where the food must reach at least 74°C at the geometric centre.

At this stage, hazards can be eliminated, provided they are properly monitored.

Cooking is the stage in which food must reach at least 74°C at its geometric centre or combinations of time and temperature such as: 65°C for 15 minutes or 70°C for 2 minutes (RDC 216).

-Defrosting

As stated in ORDINANCE 78 /RS (RIO GRANDE DO SUL, 2009), in article 9.16. Temperature of the food prepared in the cooling process reduced from 60° C to 10° C in a maximum of two hours.

4.6.4 Distribution

According to Silva Filho (1996) and Teixeira (2006), the concept of **self-service: a** system whereby the customer chooses their food directly from the plate in the quantity desired.

State health legislation provides guidance:

> 9.26. Keeping samples (100 g/100 mL) of all prepared food, including drinks (lOOmL), in suitable first-use food packaging, labelled with at least the name and date of preparation, stored for 72 hours under refrigeration, at a temperature of less than 5° C, in industrial kitchens, hotels, schools, long-stay institutions for the elderly and early childhood education establishments and other establishments (RIO GRANDE DO SUL, 2009, p. 39).

4.7 Employee training

With regard to employee training, the GRSA Environmental Management Manual (2013) defines that the Manager must ensure that the employees under his supervision receive information, training, using internal training documentation, enabling this knowledge and training to enable the operational team to influence the fulfilment of environmental, political and strategic objectives.

According to CNTL (2003), the methodology for implementing P+L must include the formation of an ecotime. In the UAN under study, this is represented by the safety team, a group of employees designed to meet and discuss issues relevant to the continuous improvement of the production process, as described in the Good Practices Manual (GRSA, 2015).

It refers to the State Health Legislation, in the item on training:

> 7.11. Handlers supervised and trained periodically (at least annually) in personal hygiene, food handling and food-borne illnesses.
> 7.12. Training proven through documentation.
> 7.13. Handlers trained on admission, covering at least the following topics: food contamination, food-borne diseases, hygienic food handling and Good Practices in food services.
> 7.14. Handlers of food services for events, mini-markets and supermarkets, street vendors and market traders who prepare and/or handle high-risk food, kitchens of long-term care institutions for the elderly, educational institutions and other places that handle high-risk food who are demonstrably trained in Good Practices (RIO GRANDE DO SUL, 2009, p. 38).

For them to learn, the content must be transmitted in a dynamic and enjoyable way, so that the employee understands and can reproduce the theoretical actions in the workplace (ABREU; SPINELLI, 2009).

Therefore, the role of the nutritionist in industrial restaurants is not limited to feeding the customer "[...] to this end, it is their responsibility to train the entire production team". (CARVALHO, 2011).

4.8 Environmental management system

The mass catering segment has sought to integrate environmental management into its business objectives. Customers are demanding that their suppliers demonstrate their commitment to achieving better environmental performance; in this climate, the demand for proven environmental responsibility is spreading throughout the supply chain, including even small companies (GRSA, 2013).

Created in 1996 by the International *Organisation for Standardisation* (ISO), ISO 14001 is an environmental management standard that aims to implement and improve an organisation's EMS, allowing its environmental performance to be maintained and potentially improved. Its basic principle is the reduction of environmental aggressors, i.e. the reduction of pollution (ISO, 1996).

The ISO 14001:2004 environmental management standard aims to provide organisations with elements of an effective EMS that can be integrated with other management requirements and help them achieve their environmental and economic objectives. The ISO 14001:2004 standard is the only one in the series developed with the aim of providing the requirements for companies to implement and obtain certification for their EMSs. This standard does not establish specific performance criteria or levels of maturity of environmental processes, nor does it establish values for control indicators, but by means of requirements the organisation must develop and implement a policy and objectives relating to significant environmental aspects (ISO 14001, 2004;

CHAVAN, 2005; CASTRO et al., 2005).

ISO 14001 requires companies to commit to pollution prevention and continual improvement as part of the normal business management cycle. The standard is based on the PDCA cycle (*plan, do, check* and *act*) and uses familiar management terminology and language. It must be appropriate to the nature, scale and environmental impacts of the organisation and includes: commitment to continual improvement, pollution prevention and legal requirements, among others. The standard must also be documented, communicated to employees and available to the public (CARVALHO, 2011).

Thus, GRSA (2013) defines that an environmental management system anchored in continuous improvement must necessarily:

- whenever possible, evaluate water and energy consumption and the level of waste produced, keeping records to identify trends and opportunities for correction/intervention;
- regularly evaluate the results and analyse, reflecting on ways to improve;
- record suggestions at regular meetings;
- allow everyone to be involved and contribute ideas;
- see problems as opportunities;
- Incidents are opportunities for learning, improvement and correction;
- set targets and monitor performance, and,
- keeping up to date with new technologies and best practices.

4.9 Environmental aspects and impacts

ISO14001 (2004) defines an environmental aspect as an "element of an organisation's activities, products and services that can interact with the environment". A note to this definition adds that "a significant environmental aspect is an environmental aspect that has or may have a significant impact on the environment". The organisation must establish and maintain an up-to-date procedure for identifying the environmental aspects of its activities, products and services. One way of identifying them is to work from regulatory and legal requirements or from legal and business risks that affect the organisation's activities. Government regulations already reflect key environmental aspects of industrial activity.

GRSA (2013) defines an environmental aspect (cause) as the element of an organisation's activities, products or services that can interact with the environment.

Another way, as mentioned in the ISO 14004 guidance standard, is to focus on products and services that create some change, whether positive or negative, in the environment. Obvious choices would include activities that result in pollution or contamination of air, water, land, solid waste or any use of raw materials or other natural resources. Companies can also focus on the results of any environmental risk assessment, any information, if available, on previous environmental incidents and all existing environmental management practices. The exercise of identifying environmental aspects encourages employees to pay attention to environmental issues that they might not have considered before.

CNTL (2003) says that this analysis can involve areas of organisational operations that may not have been considered "environmental", such as research and development (environmental considerations in product design), purchasing (evaluation of alternative raw materials) or office operations (building maintenance and recycling). There are also environmental aspects when planning new construction products or remodelling existing facilities.

According to CNTL (2003):

> The purpose of identifying environmental aspects is to determine which of them have or could have significant environmental impacts. This ensures that the aspects relating to these significant impacts are reflected in the company's objectives and targets. Identifying environmental aspects is an ongoing process, and requires organisations to keep the information up to date. The next step is to examine, evaluate and prioritise the significant environmental impacts associated with the environmental aspects of activities, products or services.

Impacts are defined in ISO-14001 (2004) as "any change in the environment, whether adverse or beneficial, wholly or partially resulting from the organisation's activities, products or services".

According to GRSA (2013), environmental impact (effect) is any change to the environment, whether adverse or beneficial, total or partial, resulting from an organisation's activities, products or services. For each environmental aspect of activities, there are one or more impacts on the environment (cause and effect).

For the production of meals, significant environmental aspects are considered: electricity consumption, solid waste disposal, water consumption and reducing the risk of physical, chemical and microbiological contamination of food (GRSA, 2013 p. 2), according to the Environmental Aspects and Impacts Worksheet (Appendix C).

CHAPTER 5

RESEARCH METHODOLOGY

5.1 Research design

The topic addressed in this work is the minimisation of waste in the food and nutrition unit using P+L concepts, seeking rational and sustainable solutions for the production of meals. It is defined as a study of alternatives for carrying out new processes that minimise harm to the environment, optimising the production process.

5.2 Unit of analysis

5.2.7 *Location*

The research work was carried out in the Food and Nutrition Unit of the service provider company, in partnership with the client company, located in the municipality of Porto Alegre, as mentioned in the Consent Form (APPENDIX A). This company was chosen because the client has environmental certification and has this as a value in its organisation. And this production process meets the requirements of the relevant health legislation.

5.2.8 *Service capacity*

The research was carried out in a *self-service* UAN, as shown in the floor plan (APPENDIX D), with only protein portioning and which serves an average of 300 customers every day at lunchtime from 8am to 2pm. The menu, as shown in Appendix B, is characterised as medium-priced and has 3 protein options, 2 garnishes, 5 types of salad, and dessert: fruit, jam and/or jelly. The establishment also has a *coffee-break* service, but the study only focused on the lunch components.

53 Data collection

The data collection procedures for this work are described below:

53.1 *Analysing documents*

The documentation relating to the UAN's production process and its waste generation over the last 18 months was analysed.

The methodological sequence is presented below:
a) Description of food and nutrition unit processes;
b) Concept of the types of waste generated;
  - Unused Product: this is defined as raw materials purchased and not used within their life cycle. For various reasons, such as: changes to the menu at the customer's request, problems with equipment, changes in the volume of services, impossibility of use due to labour difficulties.

According to GRSA (2015), an obvious opportunity to reduce waste in a facility is to avoid receiving food that cannot be consumed, based on checking all the necessary information so that it is received properly.
  - Expired: product not used within its expiry date and must be discarded immediately, subject to a health offence.

As described by GRSA (2015) for any type of raw material, in any storage condition, the criterion for physical control should be adopted, where the first to expire is the first to leave (PVPS). This item will be checked using the product's expiry date. Expiry dates must be checked in order to avoid wastage and keeping expired products in stock. Food that has deteriorated or expired must be promptly discarded, as well as food that shows alterations to the packaging or product. The presence of expired products in stock or

delivered to consumers is a health offence and a crime against consumer relations.
  - Clean leftovers: ready-to-eat food that has not been distributed, including buffets, *passthrough* ovens (intermediate storage equipment) and bain-marie ovens.
  - Pre-preparation and preparation (edible and inedible): waste generated during the preparation of each day's menu, in the areas of pre-preparation and preparation of salads, desserts, cooking and butchery: leaves, peelings, vegetable and fruit stalks that will not be used in the preparations, remains of the containers used for cooking (CARVALHO 2011).

Frying oil: "oil used [...] in frying preparations and which should be discarded". (CARVALHO, 2011, p. 37).

According to GRSA (2013, p. 1, module 3), one litre of oil can contaminate around 1 million litres of water (enough for one person to consume for 14 years).

Thus, according to the Brazilian Association of Technical Standards - ABNT (2004) in its NBR - 10.004, used lubricating oil is classified as hazardous waste, Class I and restaurant waste (food waste) is in Class II, non-hazardous waste. However, there is no clear classification for frying oils. Assessing their impacts, it can be said that they are hazardous waste, but the legislation does not classify them in this category (BOTARIO, 2009).

The oil should not be disposed of in the public sewage system; it should be stored in its own containers and contacted by companies, bodies or organisations licensed by the competent environmental agency. As employees in Food and Nutrition Units work with the frying process on a daily basis, they often need to be better informed of where and how a process can be modified to minimise the generation of waste and the emission of pollutants (BOTARO, 2009).

The literature does not record any research describing the quantity of frying oil disposed of in IRs. "[...] In the restaurant studied, the frying oil is used for preparation three times and, immediately afterwards, it is disposed of in gallons labelled with the name of the product, the sector of the IR that generated this waste, the date and origin of the product, and stored in the cold room of the rubbish bin." (CARVALHO, 2011, p. 54).

- Coffee grounds: published data on the reuse of coffee varieties is scarce. With a view to extracting value-added compounds, bibliographies have shown that this coffee residue is a good source of antioxidants and dietary fibres and could be considered a new functional ingredient (CRUZ, 2014);

c) The amount of waste generated by the unit was monitored and measured by weighing it daily in a specific container over a six-month period, from October 2015 to March 2016;

d) collection time: during the day's menu preparation (approximately 7am to 5pm);

e) each collection point had a specific container with a capacity of 14kg;

f) Waste was measured on an excel spreadsheet to be filled in manually;

g) The data was consolidated, monitored and analysed.

Waste generated at other points in the process is not included in this study, but the design covers the five points described above.

54 Materials

a) Menu: described weekly, with the preparations served to the client, constructed with information from the supplier x client contract, nutritional needs of the target audience, cost assessment, information on the number and name of the dishes to be served, planning of stages prior to the day it was served, sizing of equipment and labour available for its production;

b) daily requisition form: description of recipes with per capita data and total volume of items needed to produce a given recipe, monitoring of information on items offered more or less, depending on the characteristics of the day's stock;

c) platform scale, 300kg capacity;

d) stock item entry and exit controls: procedure where the stock team provides the quantity of each preparation for the next day's menu the day before;

e) training applied daily/weekly and monthly to the team: daily training materials received by the unit from the company's corporate areas on food safety, good handling practices, the environment, occupational safety, customer service, interpersonal relationships and other subjects aimed at training and qualifying food handlers;

f) waste monitoring spreadsheet;

g) measuring container, figure 7.

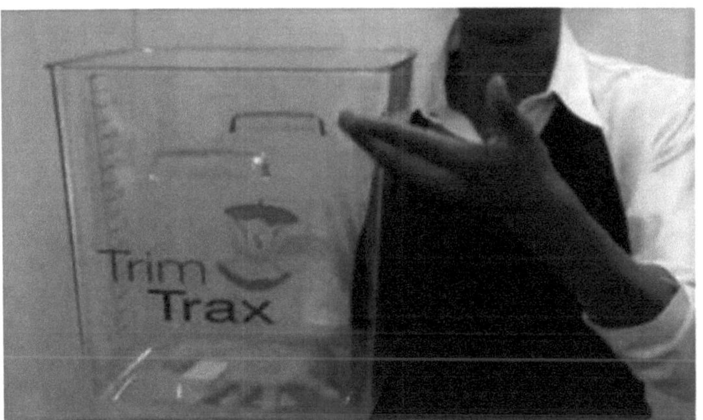

Figure 7 - Collection container

Source: GRSA, 2013.

Figure 8 describes the production process and the collection points.

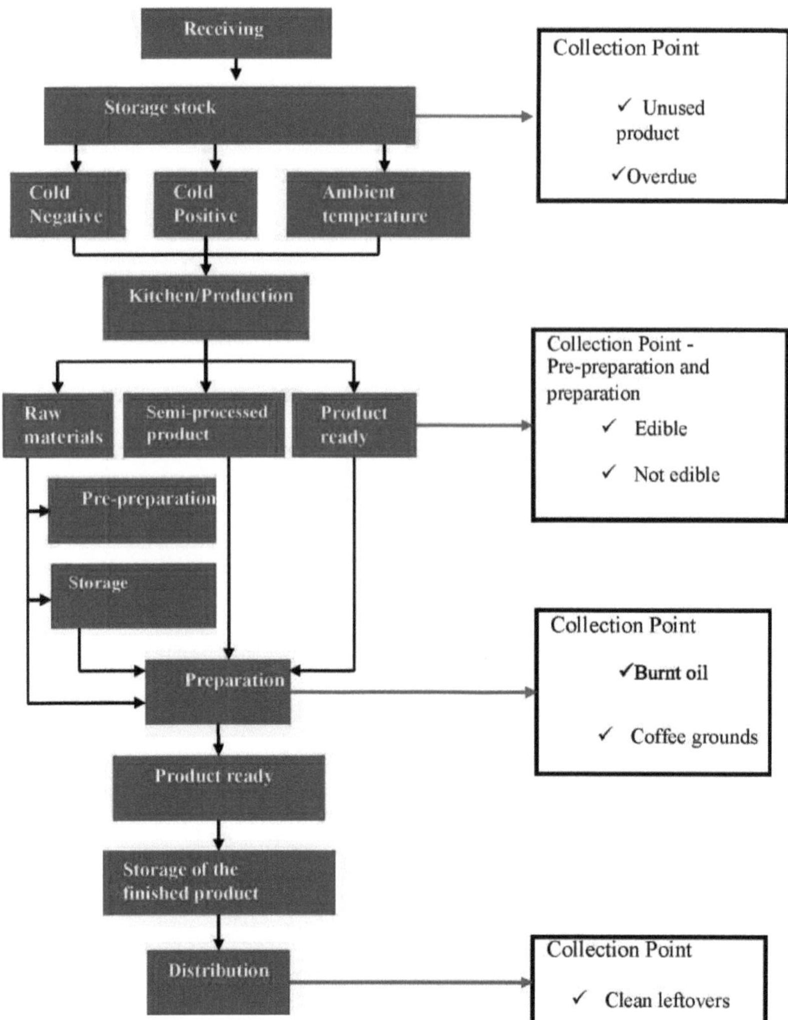

Figure 8 - Location of the collection points in the linear production flowchart
Source: GRSA, 2015.

The problems encountered during the research will be dealt with in a separate chapter.

CHAPTER 6

RESULTS AND DISCUSSION

The UAN under study has been described according to the stages of its production process and provides meals for approximately 300 people, including users of the client company and employees of the restaurant.

This research involved the restaurant building (operated by a contracted company) and its staff (nutritionist, administrative assistant, stockist, head chef, cook, kitchen officer, kitchen assistants and attendants) and the quality department of the contracted company.

The UAN operates during business hours, serving lunch from Monday to Friday. Approximately 300 meals are served per day.

6.1 Physical structure

The physical structure of the UAN under study has:

a) raw materials receiving area: where raw materials are received and checked, with access to stock and the administrative room;

b) stock room: an area where dry and perishable goods are stored. It has shelves and racks for storing products in stock and fridges and freezers for storing chilled and frozen raw materials. This area is air-conditioned for better storage and maintenance of products;

c) process area: This is where the hot and cold chain of the menu served is pre-prepared and prepared. It has worktops, sinks, a cooker, an oven, a blender, a mixer, a food processor, a fryer and an exhaust system. This is also where the intermediate storage is located, represented by the presence of the cold and hot *passthroughs*;

d) Meat area: a place next to the process area, isolated, separated by a physical barrier and properly air-conditioned, used for the pre-preparation of the protein items that make up the menu, an area used only for this purpose, in order to reduce the risk of cross-contamination;

e) Sanitisation room: made up of two rooms, the first used for sanitising utensils used in food preparation, such as pans, basins, jugs, ladle-type utensils, spoons, tongs, vats and lids. Next door is a second room for sanitising service utensils taken over from customer service. These include trays, plates and cutlery in general;

f) service area: the place where the final meal is distributed and where the public have their meal. It consists of entrance and exit turnstiles to control the public, hot and cold serving counters, support counters and sauce counters, refreshers, drinking fountains, coffee tables, tables and chairs in general;

g) toilets and changing rooms: located at the opposite end of the reception area; this area is for kitchen staff and is used for personal hygiene, storing personal belongings and changing clothes;

h) Nutritionist's room: next to the stock room, the room contains an administrative area with tables and chairs, cupboards for storing documents and a workstation with a computer, telephone and Internet access;

i) gas plant: located 100 metres away from the user's exit from the building, the site is used to store liquefied petroleum gas used in the UAN's production process;

j) waste area: located 150 metres from the restaurant building, the area contains all the waste generated by the UAN.

62 Routine activities

The operations carried out in a UAN include the hygiene and sanitation requirements of the facilities, the maintenance and sanitisation of equipment and utensils, control of the potability of water and urban pests, various training courses, control of the hygiene and health of operators, waste management and quality control of the food produced (CARVALHO, 2011).

The basic activities in the routine of a UAN are listed below:

<u>- Preparing and planning menus</u>

a) The menu is prepared according to demand;
b) raw materials are brought in on pre-defined dates to fulfil the menu.

• <u>Pre-operation</u>

Receipt of raw materials, qualitative and quantitative checks, stocking according to the type of goods.

• <u>Production</u>

Pre-preparation and cooking, which involve the requirements of health legislation.

• <u>Post-operation</u>

Distribution and secondary processes (waste disposal, cleaning and sanitising the environment and utensils).

Calculation of results, accounting for production and sales (CARVALHO 2011).

63 Functional chart

Operational unit made up of 11 employees:

- 01 Nutritionist: responsible for the general management of the unit, responsible for technical, financial and administrative processes;
- 01 administrative assistant: supports unit management in the day-to-day running of the unit, responsible for administrative processes and for monitoring and controlling documents and payroll;
- 01 attendant: employee responsible for customer service and for cleaning and organising the service area;
- 01 head chef: professional responsible for supervising and coordinating the stock, preparation, service and hygiene areas of the unit and its areas;
- 01 cook: professional responsible for producing the menu and preparing the hot chain items;
- 03 kitchen officers: support staff, responsible for the cold chain of the menu, i.e. salads, desserts;
- 03 kitchen helpers: employees responsible for cleaning and sanitising all process areas, sanitising serving and preparation utensils.

6 A **Contracted service**

The relationship between Contractor and Contractor is based on a commercial contract describing responsibilities, rights and duties. This document establishes that the contract is valid for 36 months, with an annual readjustment to be carried out in January. Billing takes place monthly, on the last working day of the month.

It is established that the building structure, i.e. construction and maintenance, is the responsibility of the contractor, as is the payment of water and electricity consumption. Waste generated by the restaurant and sanitary maintenance, such as sanitising the exhaust systems and septic tank, will be managed by the contractor.

The contractor is responsible for purchasing and maintaining equipment, supplying LPG gas, contracting and monitoring water potability, pest control, disinsection and deratisation. They must also provide telephone and internet services.

The contracted company is responsible for: supplying the raw materials for the service, managing all the services requested, managing the restaurant's workforce, purchasing uniforms and personal and collective protection equipment, and complying with health legislation requirements.

The contract describes the format of the service, the range of dishes and the themed events to be held throughout the year.

63 Public served

The public they serve is made up of administrative staff, mostly male, with technical and higher education levels. Their working hours are Mondays to Fridays during office hours.

6.6 Employee training

With regard to employee training, the GRSA Environmental Management Manual (2013) states that the manager must ensure that the employees under his supervision receive information and training using internal training documentation, enabling this knowledge and training to enable the operational team to influence the fulfilment of environmental, political and strategic objectives.

According to CNTL (2003), the methodology for implementing P+L must include the formation of an ecotime. In the UAN under study, this is represented by the safety team, a group of employees designed to meet and discuss issues relevant to the continuous improvement of the production process, as described in the Good Practices Manual (GRSA, 2015).

It refers to the State Health Legislation, in the item on training:

> 7.11. Handlers supervised and trained periodically (at least annually) in personal hygiene, food handling and food-borne illnesses.
> 7.12. Training proven through documentation.
> 7.13. Handlers trained on admission, covering at least the following topics: food contamination, food-borne diseases, hygienic food handling and Good Practices in food services.
> 7.14. Handlers of food services for events, mini-markets and supermarkets, street vendors and market traders who prepare and/or handle high-risk food, kitchens of long-term care institutions for the elderly, educational institutions and other places that handle high-risk food who are demonstrably trained in Good Practices (RIO GRANDE DO SUL, 2009, p. 38).

For them to learn, the content must be transmitted in a dynamic and enjoyable way, so that the employee understands and can reproduce the theoretical actions in the workplace (ABREU; SPINELLI, 2009).

Therefore, the role of the nutritionist in industrial restaurants is not limited to feeding the customer "[...] to this end, it is their responsibility to train the entire production team". (CARVALHO, 2011).

6.7 Waste disposal

Gathering the data obtained during the research was fundamental to identifying opportunities to minimise waste, assessing opportunities to modify processes and proposing good practices.

The waste generated at the collection points is stored in a specific container for organic waste, located in an area within the client company's plant. The area has physical barriers for pest control and this place is for the exclusive use of waste disposal, according to the protocol of the UAN studied.

Organic waste is collected three times a week by a contracted company and sent to the municipal landfill.

The burnt oil is stored in a blue vat, labelled with the collecting company and with a containment dyke, as required by environmental legislation. It is collected every fortnight.

According to studies by CNTL (2003), a series of potential barriers have been identified which can prevent or slow down the adoption of P+L in companies, including: lack of perception of the potential positive role of the company in solving environmental problems, limited scope of environmental actions within the company, among others.

63 Analysing the block diagram

According to CNTL (2003), the diagnosis of inputs (water, energy, raw materials and other inputs) and outputs (waste, atmospheric emissions, by-products and products) at each data collection point is important for analysing the aspects and impacts generated.The results identified in these diagrams are described below (Table 2 to 6).Silva (2014) reports that the mass diagram is used to summarise the identification of raw materials, waste generated and liquid effluents.

Table 2 - Unused/unexpired qualitative mass diagram

Qualitative Mass Diagram		
ENTRY	STAGE	EXIT
Raw materials: frozen products, chilled products, dry stocks, chemicals **Inputs:** water and electricity	COLLECTION POINT **UNUSED / EXPIRED**	**Waste generated:** cardboard, paper, plastics, organic **waste** in general, dry **waste**, office paper, chemicals **Effluents:** sewage

Source: authorship, 2016.

Table 3 - Qualitative mass diagram pre-preparation/preparation

Qualitative Mass Diagram		
ENTRY	STAGE	EXIT
Raw materials: perishable products, proteins in general, fruit and vegetables, dry stocks **Inputs:** water and electricity	COLLECTION POINT **PRE-PREPARATION/PREPARATION**	**Waste generated:** plastic, plastic and paper labels, organic in general, dry **Effluents:** sewage

Source: authorship, 2016.

Table 4 - Qualitative mass diagram for burnt oil

Qualitative Mass Diagram		
ENTRY	STAGE	EXIT
Raw materials: perishable, frozen, chilled and stock products **Inputs:** water and electricity	COLLECTION POINT **BURNING OIL**	**Waste generated:** burnt oil and organic matter **Effluents:** sewage

Source: authorship, 2016.

Table 5 - Qualitative mass diagram of coffee grounds

Qualitative Mass Diagram		
ENTRY	STAGE	EXIT
Raw materials: coffee powder **Inputs:** water and electricity	COLLECTION POINT **COFFEE BUTTER**	**Waste generated:** coffee grounds

		Effluents: sewage

Source: authorship, 2016.

Table 6 - Qualitative mass diagram of clean leftovers

Qualitative Mass Diagram		
ENTRY	STAGE	EXIT
Raw materials: products ready for distribution and consumption **Inputs:** water and electricity	**COLLECTION POINT** **CLEAN SOIL**	**Waste generated:** organic waste, PVC plastic (vat protection) **Effluents:** sewage

Source: authorship, 2016.

6.9 Profile of the waste generated at each stage between October 2015 and March 2016

Figure 8 shows the evolution of waste - the target of this study. Over one tonne of waste was generated during the period, with pre-preparation and preparation being the most significant area.

In the other areas, data was generated:

J 90 kg - Unused/ expired

J 82 kg - Coffee grounds

■*S* 395 kg - Clean leftovers

372 litres - Frying oil

The analysis of the data obtained, which outlines the profile shown in figure 9, suggests that raising staff awareness, opportunities to minimise waste generation at source, process changes and the adoption of good practices are all actions that led to the decision to invest in P+L.

It is therefore possible to minimise waste generation, seek efficiency in the use of natural resources and generate sustainable improvements in the long term (CNTL, 2003).

The pre-preparation stage produces the largest amount of waste by mass, accounting for 47 per cent of the volume collected during the survey.

Of equal importance, but with less impact, burnt oil is a factor to be considered. Greater control in the use of the fryer, critical analysis of the use of fried food, are actions to be reviewed by the process manager with economic and environmental benefits "[...] as a cost of poor quality, we cite equipment costs, whether in poor condition, inefficient, underutilised or used inappropriately". (ABREU, 2009, p. 39).

Lack of management or inadequate management and the data collected from clean leftovers require an in-depth analysis of menu planning and the user's consumption profile. According to Abreu (2009), control consists of comparing the execution with the planning of the objectives set and the results obtained.

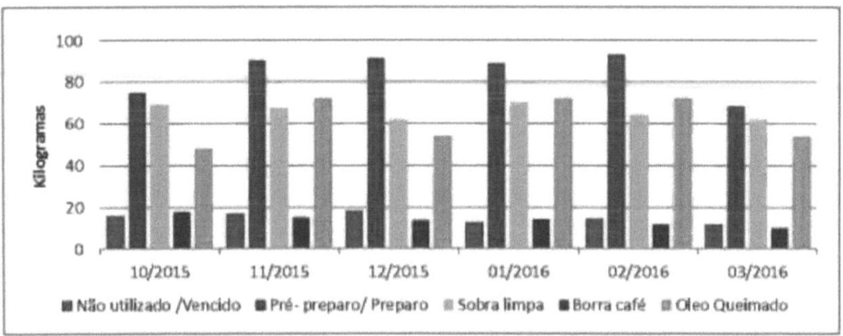

Figure 9 - Profile of the waste generated at each stage between October 2015 and March 2016

Source: authorship, 2016.

However, the results obtained in this study indicate that actions to intensify management and change the behaviour of the team of workers are important factors in reversing this scenario, as stated by CNTL (2003, p. 39) "[...] cleaner production focuses on a strategy to continuously reduce pollution and environmental impacts through reductions at source - eliminating waste within the process rather than treating it after it has been generated".

6.9.1 Unused and expired waste

The data in figure 9 on unused/ expired waste obtained in this research shows that the raw material storage and management stage is vital to any production process. As the data collected shows, there are numerous benefits to correctly managing inputs. Such as:

- purchase adjusted to real production needs;

- evaluating the seasonality of products, seeking cost-benefit ratios;

- adjusted product shelf life control;

- proper sizing between product acquisition and storage capacity.

According to ABREU (2009, p. 119), "Errors in the management of these logistics translate into irregular replacement of raw materials, larger quantities of stock without changes in consumption, lack of storage space [...]."

There were barriers such as the high turnover of professionals in this area and low commitment to data monitoring.

41

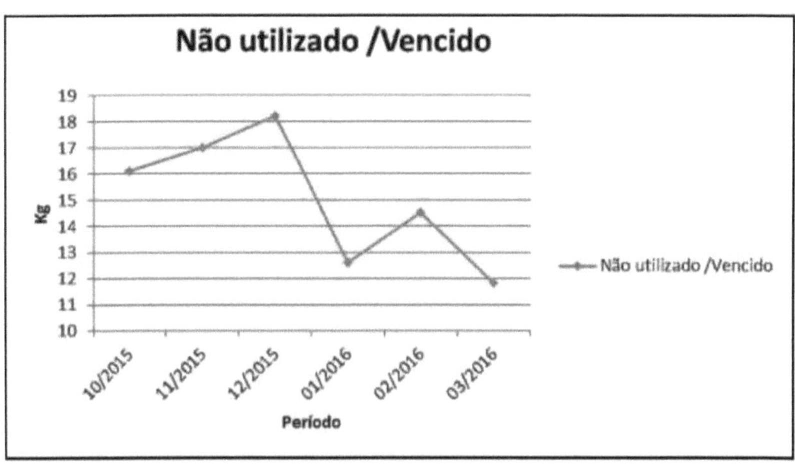

Figure 10 - Unused/overdue collection point from October 2015 to March 2016
Source: authorship, 2016.

Figure 10 shows the improvements achieved with input management and staff training. There was a reduction from 18kg/month in December to 12kg/month in 01/2016, as a result of the actions established in the UAN and mentioned below with suggestions for P+L improvements.

The causes of waste at this stage include a lack of organisation of spaces (shelves, monoblocs); products without identification or with incomplete identification, leading to doubts as to the expiry date; lack of cleaning; expired products; the presence of insects and rodents and insufficient maintenance of equipment temperatures (BRADACZ, 2003).

Suggestions for improving P+L

P+L level 1 - reduction at source

- Focus on training and capacity building for employees in this area;
- changing the planned menu to utilise raw materials that are nearing the end of their life cycle;
- changing the delivery of fruit and vegetable orders to daily;
- weekly supervision of the area by the Nutritionist, with the aim of helping to control the shelf life of the products;
- preventive maintenance of cooling equipment, with the aim of keeping it at the correct temperature parameter, helping with the life cycle of products;
- equipment sizing; with the acquisition of equipment for storing **fresh** fruit **and** vegetables.

6.9.2 Edible and inedible pre-preparation and preparation waste

The data in figure 11 on pre-preparation and preparation obtained in this research shows a 26 per cent variation in waste generation. The P+L suggestions mentioned below were decisive for the reduction found at the end of the period, where the monthly figure was below 70 kg/month.

In January and February 2016, *fresh* raw materials were used to prepare salads and side dishes, due to the high cost of pre-processed foods.

In March, items such as potatoes, carrots, cabbage and chayote were purchased with processed products, a reduction of 21 per cent.

In this study, the per capita obtained at the pre-preparation stage between October and December 2015 was 0.013/kg

In Carvalho (2011), during the same period of the year, the per capita resulting from the waste in the pre-preparation stage was, on average, 0.018 kg.

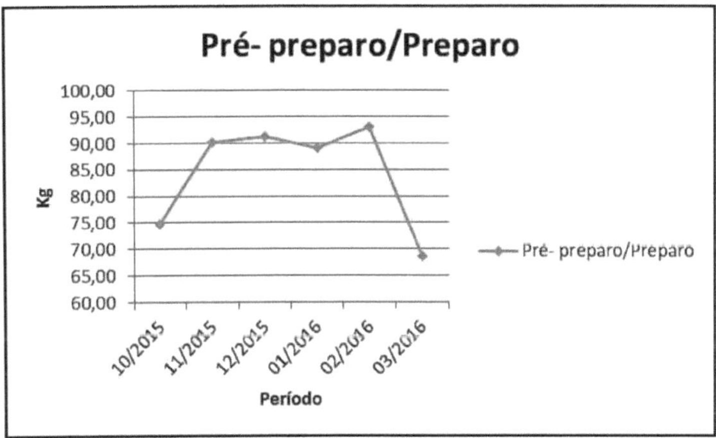

Figure 11 - Pre-preparation collection point / edible and non-edible preparation during the months of October 2015 to March 2016

Source: authorship, 2016.

According to BOTARO (2011), it is important to emphasise that the generation of waste in pre-preparation and preparation is significantly influenced by the training of the employee responsible for this stage.

Adequate work tools and specific training are determining factors in the fall in the value expressed in 03/2016. Continuous use tends to keep waste generation to a minimum.

Suggestions for improving P+L

P+L level: 1 - reduction at source

43

1. Purchase of processed fruit and vegetable items, especially leafy greens;

2. purchasing proteins, especially beef, in portioned form. Avoiding cutting the meat in the unit, seeking greater utilisation of the raw material purchased;

3. planning the Monday menu, trying to include preparations with a low percentage of pre-preparation;

4. evaluate losses due to protein melting;

5. provide the employee with a vegetable peeler, a tool that improves performance in tasks such as peeling items in order to maximise the use of the input.

6.9.3 Burnt oil residue

The data in figure 12, on burnt oil waste, obtained in this research shows the consumption of oil used in the fryer during the period, where the value found was 10 ml per capita.

According to the literature, the total amount of oil used in the preparation of the menu can vary between 17 ml and 21 ml (AMORIN; JUNQUEIRA; JOKL, 2010).

During the study period, figure 12 shows that the assessment of the quality and condition of the oil used in the fryer was not standardised.

According to Ordinance 78 (RIO GRANDE DO SUL, 2009, p. 38):

> 9.7 Oils and fats used heated to a temperature not exceeding 180° C.9.8 Oils and fats replaced when there is an obvious change in their physical-chemical or sensory characteristics (smoke, foam, aroma and flavour).9.9 Monitoring the quality of oils and fats for frying with records of this control.

According to 3M (2016), oil degradation percentages indicate its degree of saturation:

- 2 per cent of the oil began to degrade;

- up to 3.5% use the oil for more sensitive foods such as potatoes and polenta;

- from 3.5% to 5.5% use this oil for tougher foods such as breaded and;

- from 5.5% to 7% do not use for any food.

In 12/2015, the fryer was used twice a month, resulting in more than 50 litres/month. On the other hand, in 03/2016, there was a 100% increase in incidents of preparation using the fryer, compared to 12/2015, with adequate management and monitoring, where the waste generated in the month was equal to the value found in 12/2015. This shows the contribution of the suggestions to minimising waste generation.

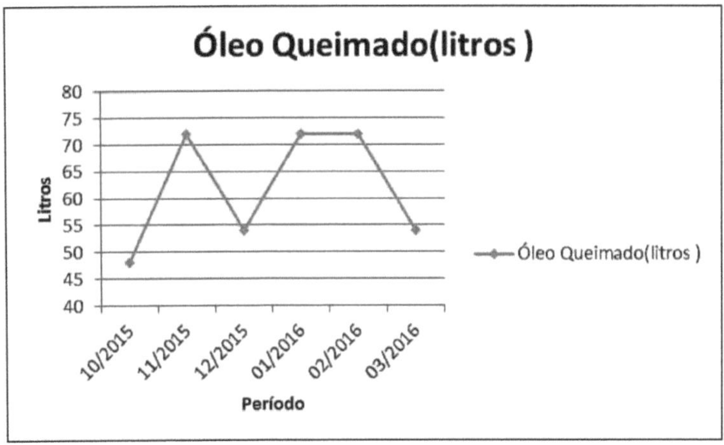

Figure 12- Burning oil collection point from October 2015 to March 2016

Source: authorship, 2016.

Suggestions for improving P+L

P+L level: 1 - reduction at source

1. Training for employees to enable them to identify indicators of saturation, sensory quality and odour of burnt oil;

2. Implementation of oil quality monitoring, using an oil saturation measuring tape;

3. Menu planning with a focus on the degree of saturation of the preparations.

6.9.4   Coffee grounds waste

The data in figure 13 on coffee grounds waste obtained in this research shows the potential of the waste generated at this collection point for use at level 3 of the P+L actions, which focus on opportunities to reuse waste, with external recycling. As shown in figure 14, after the company was made aware of the P+L programme, there was a reduction in the waste generated by making some changes to the process, without compromising the quality of the end product. Month on month, the generation of waste has decreased, with the P+L suggestions presented.

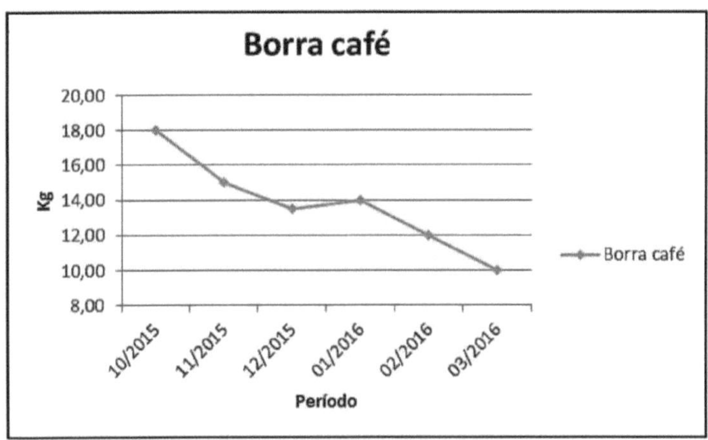

Figure 13 - Coffee grounds collection point from October 2015 to March 2016

Source: authorship, 2016.

According to CNTL (2003, p. 29) good operating practices include [...] changing the dosage and concentration of products.

According to Cabral and Moris (2010), the study of the application of oil obtained from the extraction of coffee grounds in biodiesel focuses on the reuse of waste and also uses coffee grounds as a reinforcing agent in polymeric films. Reusing waste adds economic value to products, by-products and waste from production processes, reducing impacts on the environment, encouraging the non-generation of waste, the recycling of raw materials and/or by-products and avoiding the generation of environmental liabilities.

Suggestions for improving P+L

P+L level: 1 reduction at source

1. Implementation of a technical sheet to standardise the coffee/litre recipe;
2. carry out a customer satisfaction survey in order to understand why the product has not been consumed.

P+L level 3 - external recycling

1 Develop a partnership with an external company to send this waste for reuse.

6.9.5 Clean leftover waste

The data in figure 13 on coffee grounds waste obtained in this research shows opportunities for improvement in the clean leftovers stage, based on production monitoring and the consumption profile of the end customer. The results achieved in the months of 12/2015 and 03/2016 stand out, where actions to raise awareness among the team and management were intense, with the aim of organising the production process to produce in batches, according to the demand reported by the service team.

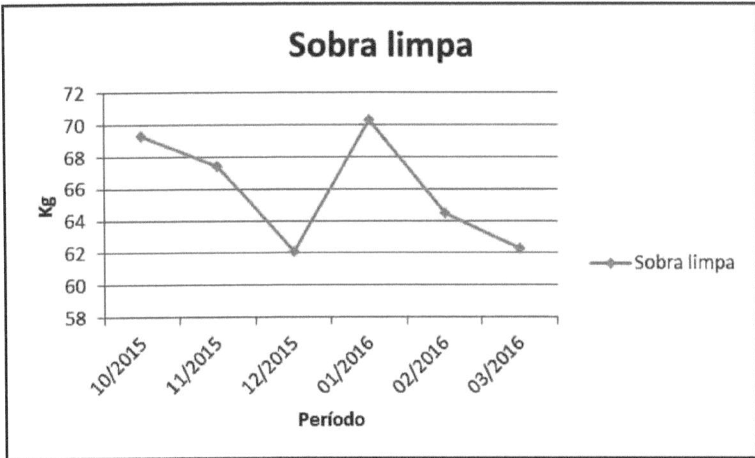

Figure 14 - Clean leftovers collection point from October 2015 to March 2016

Source: authorship, 2016.

A restaurant has a complex routine, with a large number of daily activities that must be carried out simultaneously within a given timeframe. If clean leftovers are seen as a detail, they can increase the generation of waste, increase operating costs and jeopardise the bottom line (CARVALHO, 2011).

Suggestions for improving P+L

P+L level: 1 reduction at source

1. With the exception of beans, the other preparations should be produced in small batches in order to monitor production x consumption;

2. evaluate the quantity to be produced, such as the result of last week's consumption, with the aim of fine-tuning the number of services served in recent days;

3. parameterisation of the operating temperature of the *passthrough* so that the equipment keeps the preparation in good condition until the end of the service, without the need to produce a new batch due to changes in the dish while in storage,

4. to plan the production of the daily menu based on the profile of the day of the week, it was observed that the user migrates to an external location on Fridays.

It was noted that of the 21 suggested improvements, most of which were classified as level 1, 4 were implemented.

This result amply demonstrates the possibility of reducing environmental impacts in the production process. Even considering that some measures require investment, some of them can be done at low cost.

Environmental awareness and the collaboration of those involved represents a major problem.

The involvement of the manager, in this case the nutritionist, in the process of changing the vision of employees to break paradigms, is fundamental, even if it is continuous and slow.

According to CNTL (2003), P+L initially provides opportunities for level 1 and 2 actions. When opportunities for applying levels 1 and 2 are extinguished, level 3 opportunities will be evaluated.

Actions to minimise the waste generated have been implemented, but a large percentage of it is still sent to the "end of the pipe". There is a need to change perceptions, to carefully assess the input of raw materials, the life cycle of products, to invest in raising awareness and to stop analysing waste in the traditional way.

CHAPTER 7

CONCLUSION

The results obtained show that this UAN, after characterising its production process and learning about the various aspects involved in customer service, has made progress in quantifying its waste and making its staff aware of the environmental impact generated.

At the collection points during the period, 1,075 kg were generated to serve 34,736 meals.

For the unused/expired collection point, there was a 45 per cent reduction in waste generation, comparing the month with the highest number of kilos found with the month with the lowest.

At the clean leftovers collection point, there was a 12 per cent variation in monthly waste generation. It is important to emphasise that this collection point directly reflects the financial loss of the meal produced and not invoiced.

Nutritionists should start by raising awareness about the use of natural resources, as well as critically analysing the production process in order to minimise the generation of waste.

Points of waste generation determined by the production process were identified, and knowledge of the raw material profile and monitoring and control procedures was deepened. Based on this identification, P+L concepts were applied to minimise waste.

One benefit that could be observed was the change in management and routines in the purchasing/stock and warehousing area, with clear definitions of procedures, tasks and objectives.

It should be noted that the UAN understood the importance of planning and executing its menu, seeking to minimise waste; as an action resulting from technical, organisational, conceptual and technological changes.

The good practices proposed, such as: drawing up the menu with analysis requirements, seeking greater productivity from the oil used in the fryer, and the correct disposal of coffee grounds.

Good practices support the UAN in minimising the waste generated and provide results that can be applied to other UANs.

The biggest difficulty encountered was the variety of inputs accounted for at the pre-preparation/preparation collection point (bones, rinds, trimmings, butchers, broths, shavings) combined into a single piece of data, where there was a 34% variation in the waste generated, influenced by the months in which items were purchased that had already been processed. In the future, the process area to which the raw material belongs could be subdivided in order to individualise the data collected.

There are other waste generation points on the site that were not the subject of this study.

Therefore, it can be concluded that using P+L tools is an action that in the medium and long term will provide social, environmental and economic benefits. It is also important to note that assessing the minimisation of waste generation will bring gains in productivity and costs.

7.1 Suggestions for future work

During the course of the research, it was realised that there was an opportunity to continue studying the P+L methodology in the production of meals:

To analyse in detail the waste generated in the pre-preparation and preparation stage, with the aim of expanding options and solutions.

Energy analysis of the production process, assessing possible losses and seeking greater efficiency and lower consumption of natural resources.

Evaluate water reuse possibilities within the process and quantify the savings obtained.

Evaluate the influence of technological innovation on the automation of certain processes.

To characterise the level of knowledge of employees regarding the consumption of natural resources, waste generation and environmental impacts.

REFERENCES

3 M Solutions: commercial solutions. 3M oils and fats monitor. c2016. Available at: <http://solutions.3m.com.br/wps/portal/3M/pt_BR/CuidadoInstitucional/Home/Solucoes/Foo dService/MmentoOleosGorduras/MonitorOleosGorduras/>. Accessed: 02 Feb. 2016.

ABERC. BRAZILIAN ASSOCIATION OF COLLECTIVE MEALS COMPANIES. São Paulo: ABERC, 2008. Available at: <http://www.aberc.com.br/.> Accessed on: 05 March 2016.

ABERC. BRAZILIAN ASSOCIATION OF CATERING COMPANIES. **Manual of practices for preparing and serving meals for the public.** 9. ed. São Paulo: ABERC, 2009.

ABERC. BRAZILIAN ASSOCIATION OF COLLECTIVE MEALS COMPANIES. **Real market.** São Paulo, 2016. Available at: <http://www.aberc.com.br/mercadoreal.asp?IDMenu=21>. Accessed on: 10 March 2016.

ABNT. BRAZILIAN ASSOCIATION OF TECHNICAL STANDARDS. **ABNT NBR 10151:2000: acoustics - evaluation of noise in inhabited areas, aiming at comfort - procedure.** São Paulo, 2000. 4 p.

ABNT. BRAZILIAN ASSOCIATION OF TECHNICAL STANDARDS. **ABNT NBR 10004:2004:** classification of solid waste. São Paulo, 2014. 71 p.

ABRASEL. BRAZILIAN ASSOCIATION OF BARS AND RESTAURANTS. 2012. Available at: <http://www.abrasel.com.br/>. Accessed on: 05 March 2016.

ABRASEL. BRAZILIAN ASSOCIATION OF BARS AND RESTAURANTS. 2015. Available at: <http://www.abrasel.com.br/>. Accessed on: 05 March 2016.

ABREU, E. S.; SPINELLI, M. G. N.; ZANARDI, A. M. P. **Gestão de Unidades de Alimentação e Nutrição:** um modo de fazer. São Paulo: Metha, 2003.

ABREU, E. S; SPINELLI, M. G. N. Production Evaluation. In: ABREU, E. S; SPINELLI, M. G. N; PINTO, A M. **Gestão de Unidades de Alimentação e Nutrição:** um modo de fazer. 2. ed. São Paulo: Metha, 2007. p. 205 211.

ABREU, E. S; SPINELLI, M. G. N. Production Evaluation. In: ABREU, E. S; SPINELLI, M. G. N; PINTO, A M. **Gestão de Unidades de Alimentação e Nutrição:** um modo de fazer. 3. ed. São Paulo: Metha, 2009. p. 107 -118.

ALVES, Mariana Gardim; UENO, Mariko. Identifying sources of solid waste generation in a food and nutrition unit. **Ambiagua:** na Interdisciplinary Journal of Apllied Science, Taubaté, v. 10, n. 4, p. 874-888, dec. 2015. Available at: <http://www.scielo.br/pdf/ambiagua/vl0n4/l980-993X-ambiagua-10-04-00874.pdf>. Accessed on: 01 Apr. 2016.

AMORIN, Maria Marta Amacio; JUNQUEIRA, Roberto G; JOKL, Lieselotte. Consumption of oil and fat in self-service lunch preparations. **Alimjiutr.,** Araraquara, v. 21, n. 2, p. 217-223, jan. 2010. Available at: <http://serv- bib.fcfar.unesp.br/seer/index.php/alimentos/article/viewFile/986/a7v21n2.pdf>. Accessed on: 01 May 2016.

ANVISA. NATIONAL HEALTH SURVEILLANCE AGENCY. Resolution - RDC N° 216, of 15 September 2004. Technical Regulations on Good Practices for Food Services. **Federal Official Gazette,** Brasília, DF, 16 September 2004. Available at: <http://portal.anvisa.gov.br/wps/wcm/connect/4a3b680040bf8cdd8e5dbflb0133649b/RESQL U%C3%87%C3%830- RDC+N+216+DE+15+DE+SETEMBRO+DE+2004.pdf?MQD=AJPERES>. Accessed on: Oct. 2015.

ASBRAN. BRAZILIAN NUTRITION ASSOCIATION. **History of the nutritionist in Brazil:** 1939 to 1989. Collection of statements and documents. São Paulo: Atheneu, 1991.

BILCK, A.P. et al. Making the most of of by-products: a restaurant in Londrina. **Revista em agronegócio e meio ambiente,** v. 2, n. 1, p. 87-104, jan./abr. 2009.

BOTARO, Flavia Alessandra da Silva. **Minimising Soya Oil Waste from Frying in Food and Nutrition Units.** 2009. 101 f. Thesis (PhD) - Federal University of Ouro Preto, Postgraduate Programme in Environmental Engineering, Faculty of Engineering, Ouro Preto, 2009. Available at: <http://www.repositorio.ufop.br/handle/123456789/3326>. Accessed on: 10 Feb. 2016.

BRADACZ, Dulce-Cleá. **Quality management model for food waste control in Food and Nutrition Units.** 2003. 173 f. Dissertation (Master's) - Federal University of Santa Catarina, Postgraduate Programme in Production Engineering, Faculty of Engineering, Florianópolis, 2003. Available at: <https://repositorio.ufsc.br/bitstream/handle/123456789/85188/225881.pdf?sequence=l&isAl lowed=y>. Accessed on: 01 January 2016.

BRAZIL. Constitution (1988). Constitution of the Federative Republic of Brazil of 1988. Available at: <http ://www.planalto .gov.br/ccivil_03/constituicao/constituicaocompilado .htm >. Accessed on: 05 January 2016.

BRAZIL. National Environment Council. CONAMA Resolution n. 001, of 08 March 1990. Available at: <http://www.mma.gov.br/port/conama/res/res90/res0190.html>. Accessed on: 05 March 2016.

BRAZIL. National Environmental Council. CONAMA Resolution no. 237, of 19 December 1997. Available at: < http://www.mma.gov.br/port/conama/res/res97/res23797.html>. Accessed on: 05 March 2016.

BRAZIL. National Environment Council. CONAMA Resolution no. 275, of 25 April 2001. Establishes the colour code for the different types of waste, to be adopted in the identification of collectors and transporters, as well as in information campaigns for selective collection. **Diário Oficial da União,** Brasília, 19 June

2001. Available at: <http://www.mma.gov.br/port/conama/legiabre.cfin?codlegi=273> Accessed on: 05 March 2016.

BRAZIL. National Environment Council. CONAMA Resolution n. 430, of 13 May 2011. Provides for the conditions and standards for discharging effluents, complements and amends Resolution 357 of 17 March 2005 of the National Environment Council (CONAMA). Available at: <http://www.mma.gov.br/port/conama/legiabre.cfin?codlegi=646>. Accessed on: 05 March 2016.

BRAZIL. Ministry of the Environment. **Law 6.938/81.** Provides for the National Environmental Policy, its purposes and mechanisms for formulation and application, and other measures. Available at: <http://www.planalto.gov.br/ccivil_03/leis/16938.htm>. Accessed on: 05 March 2016.

BRAZIL. Ministry of the Environment. National solid waste policy. 2010. Available at: <http://www.mma.gov.br/pol%C3%ADtica-de-res%C3%ADduos- s%C3%B31idos>. Accessed on: 05 January 2016.

CABRAL, Mariana Szente; MORIS, Virgínia Aparecida da Silva. Reuse of coffee grounds as a measure to minimise waste generation. Encontro nacional de engenharia de produção: Associação Brasileira de Engenharia de Produção, 30, São Carlos - SP. **Proceedings...** São Carlos - SP, 12 October 2010, p.1-9. Available at: <http://www.abepro.org.br/biblioteca/enegep2010_tn_stp_121_788_17072.pdf>. Accessed on: 01 Feb. 2016.

CARMO, T.V.B et al. Use of UK food waste to produce organic fertiliser for campus tree planting. **Revista brasileira de agroecologia,** v. 4, n. 2, nov. 2009.

CARVALHO, Rita de Cássia Reis. **Method for Determining Solid Waste Generation Indicators in Large Industrial Restaurants.** 2011.105 f.
Dissertation (Master's Degree) - Araraquara University Centre, Postgraduate Programme in Regional Development and Environment, Regional Dynamics and Sustainability Alternatives, Araraquara - SP, 2011. Available at:
<http://www.uniara.com.br/arquivos/file/cursos/mestrado/desenvolvimento_regional_meio_a mbiente/dissertacoes/2011/rita-de-cassia-reis-carvalho.pdf>. Accessed on: 05 Dec. 2015.

CAVALLI SB, Salay E. People management in commercial meal production units and food safety. **Revista de Nutrição,** v. 20, n. 6, p. 657-667, 2007.
CHAMBERLEM, S. R.; KINASZ, T. R.; CAMPOS, M. P. F. F. Organic waste in food and nutrition units. **Alim. Nutr.,** Araraquara, v. 23, n. 2, p. 317-325, apr./jun. 2012.

CFN. FEDERAL COUNCIL OF NUTRITIONISTS. CFN Resolution 380/2005. Provides for the definition of nutritionists' areas of activity and their attributions, establishes numerical reference parameters, by area of activity, and makes other provisions.
Available at: <http://www.cfii.org.br/novosite/pdf7res/2005/res380.pdf>. Accessed on: 05 March 2016.

CFN. FEDERAL COUNCIL OF NUTRITIONISTS. Law No. 8.234, of 17 September 1991 (DOU 18/09/1991). **Regulates the profession of nutritionist and determines other measures. [2016].** Disponível em: <http://www.cfii.org.br/index.php/lei-no-8-234-de-17- de-setembro-de-1-991-dou-18091991-2/>. Accessed on: 05 March 2016.

CHAMBERLEM, Suellen Regina; KINARSZ, Tania Regina. Leftovers and discarded leftovers - source of organic waste generation in food and nutrition units in Cuiabá-MT. **Alimjiutr,** Araraquara, v. 23, n. 2 , p. 317-325, jun. 2012. Available at: <http://serv-bib.fcfar.unesp.br/seer/index.php/alimentos/article/viewFile/2039/2039>. Accessed on: 01 May 2016.

CHIEREGATTO, C.M.P.; CLARO, J.A.C.S. Reverse logistics in commercial restaurants in the city of Santos. **Revista da micro e pequena empresa,** Campo Limpo Paulista, v. 3, n. 3, p. 96-110, 2010.

CNTL. NATIONAL CENTRE FOR CLEAN TECHNOLOGIES. **Bakeries.** Porto Alegre: Senai/Fiergs, 2005. 58 p. (P+L).

CNLT. NATIONAL CENTRE FOR CLEAN TECHNOLOGIES. **Cleaner production in bakeries and patisseries.** Porto Alegre: Senai/Fiergs, 2008.

COLARES, L.G.T. et al. **Generation of Solid waste in large-scale meal production in the state of Rio de Janeiro - BRAZIL.** 2008.

CORRÊA, Margareth da Silva; LANGE, Liséte Celina. Solid waste management in the public catering sector. **Pretexto 2011,** Belo Horizonte, p. 29-54, Mar. 2011. Available at: <www.fumec.br/revistas/pretexto/article/download/659/56>. Accessed on: 01 March 2016.

CORRÊA, M. S. The **challenge of solid waste management in food and nutrition units:** proposals for legislation, institutions and professional training. 2014. 146 f. Thesis (PhD) - Federal University of Minas Gerais, Postgraduate Programme in Sanitation, Environment and Water Resources, Belo Horizonte, 2014.

CRUZ, Rebecca. **Coffeeby-Products Sustainable Agro-Industrial Recovert and Impacton Vegetables Quality.** 2014. 117 f. Thesis (PhD) - University of Porto, Faculty of Pharmacy, Porto, 2014. Available at: <https://repositorio- aberto.up.pt/bitstream/10216/76762/2/102682.pdf>. Accessed on: 10 Mar. 2016

DIAS, Reinaldo. **Environmental Management.** 2.ed. São Paulo: Atlas, 2011.
DONADON, Kelli Cristina Padilha André. **A study on the adoption of Cleaner Production: the case of a food industry.** 2005. 102 f. Dissertation (Master's) - Araraquara University Centre, Regional Development and Environment Course, Araraquara, 2005.

DONAIRE, D. **Gestão Ambiental na Empresa.** São Paulo: Atlas, 1999.

GIANNETTI, B.F.; ALMEIDA, C.M.B.V. **Industrial Ecology:** concepts, tools and applications. São Paulo: Edgard Blucher, 2006.

GRSA. **Environmental management manual,** São Paulo, 2013.

GRSA. **Good Practices Manual.** São Paulo, 2015.

IBGE. BRAZILIAN INSTITUTE OF GEOGRAPHY AND STATISTICS. Ministry of Planning, Budget and Management. Directorate of Research, Coordination of Labour and Income. **Household Budget Survey 2008-2009:** per capita household food acquisition Brazil and Major Regions. Rio de Janeiro: 2010. Available at: <http://biblioteca.ibge.gov.br/biblioteca-catalogo?view=detalhes&id=245130>. Accessed on: April 2016.

ISO. INTERNATIONAL ORGANISATION FOR STANDARDISATION. **ISO 14001:1996:** environmental management systems. Available at: <http://www.iso.org/iso/home/standards/management-standards/iso14000.htm>. Accessed on: 05 March 2016.

ISSO. INTERNATIONAL STANDARDISATION ORGANISATION. **ISO 14001:1996:** environmental management systems: specification and guidelines for use. 1996. <http://www.labogef.iesa.ufg.br/labogef/arquivos/downloads/NBRISO14001_59064.pdf>. Accessed on: 05 Feb. 2016.

LEAL, Daniele. Growth of eating away from home. **Revista Segurança Alimentar e Nutricional,**

Campinas, v. 17, n. 1, p. 123-132, 2010.

LIMA, Luiz M. Q. **Lixo:** tratamento e biorremediação. São Paulo: Hemus, 1995.

MARINHO, Maerbal B.; KIPERSTOK, Asher. Industrial ecology and pollution prevention: a contribution to the regional debate. **Bahia Análise e Dados,** Salvador, v. 10, n. 4, p. 271- 279, mar. 2001.

MARTINS, Aline de Moraes. **Environmental Sustainability in Collective Food and Nutrition Units in Santa Catarina.** 2015. 161 p. Dissertation (master's degree) - Federal University of Santa Catarina, Florianópolis, 2015. Available at: <https://repositorio.ufsc.br/handle/123456789/135679>. Accessed on: 05 March 2016.

MARTINELLI, S.S. et al. Water Consumption in Meat Thawing under Running Water: Sustainability in Meai Production. **Journal Culinary Science & Technology,** n. 4, p. 311- 325,2012.

MEDEIROS, Caroline Opolski. **People management and food safety in commercial restaurants:** a study in Campinas, Porto Alegre and Florianópolis. 2010. 211 f. Dissertation (master's) - Federal University of Santa Catarina, Nutrition Programme, Federal University of Santa Catarina. Florianópolis, 2010.

MENDES, Fabiana Amaral Rodrigues. **Cleaner Production.** [2009]. Available at: <www.techoje.com.br/site/techoje/categoria/abrirPDF/929>. Accessed on: 07 June 2015.

MEZOMO, Iracema F. Barros. **Food services:** planning and management. 5. ed., current and revised. São Paulo: Manole, 2002.

MOURA, Luiz Antônio Abdalla de. **Quality and environmental management.** 3. ed. São Paulo: Juarez de Oliveira, 2002.

NASCIMENTO, Luis Felipe; LEMOS, Angela D.C.; MELLO, Maria C.A. **Produção mais Limpa[Rio Grande do Sul],** Universidade Federal do Rio Grande do Sul - UFRGS, 2002.1 Cd-Rom.

PHILIPPI, Sonia Tucunduva. **Nutrição e técnica dietética.** 2. ed. Barueri: Manole, 2006.

PORTO ALEGRE. Complementary Law no. 728, of 8 January 2014. Establishes the Municipal Urban Cleaning Code, repeals Supplementary Laws no. 234, of 10 October 1990, 274, of 25 March 1992, 376, of 3 June 1996, 377, of 3 June 1996, 591, of 23 April 2008, and 602, of 24 November 2008, and makes other provisions.
Available at: < http://www2.portoalegre.rs.gov.br/cgi-bin/nph-brs?s1=000033832.DOCN.&l=20&u=%2Fnetahtml%2F sirel%2Fsimples.html&p= 1 &r= 1 &f= G&d=atos&SECTl=TEXT>. Accessed on: 05 June 2015.

PROENÇA, R.P.C. et al. **Nutritional and sensory quality in the production of meals.** Florianópolis: UFSC, 2005.

RIEHL, Alice. **Minimising waste generation in a tannery using tools from the P+L programme.** 2008. 100 f. Monograph (Graduation) - Vale dos Sinos University, Faculty of Production Engineering, São Leopoldo, 2008.

RIEKES, B.H. **Quality in food and nutrition units:** a methodological proposal considering nutritional and sensory aspects. 2004. 170 f. Dissertation (master's degree) - Federal University of Santa Catarina, Postgraduate Programme in Nutrition, Florianópolis, 2004.

RIO GRANDE DO SUL. Secretary of Health of the State of Rio Grande do Sul. Ordinance No. 78/2009 of 30 January 2009. Approves the Good Practices Checklist for Food Services. Approves Standards for

Training Courses in Good Practices for Food Services and makes other provisions. **Federal Official Gazette,** Porto Alegre, 30 January 2009. Available at: <http://www.saude.rs.gov.br/upload/1365096500_portaria%2078_09.pdf>. Accessed on: 05 June 2015.

SALES, Gizene L. P. de. **Diagnosis of solid waste generation in popular public restaurants in the municipality of Rio de Janeiro: a** contribution to minimising waste. 2009. 140 f. Dissertation (Master's) - Federal University of Rio de Janeiro, Postgraduate Programme in Nutrition, Rio de Janeiro, 2009. Available at: <http://www.dominiopublico.gov.br/download/texto/cp093589.pdf>. Accessed on: 29 June 2015.

SANCHEZ, V. D. **Diagnosis of Organic Solid Waste Generation at the University Restaurant of the Universidade Estadual Paulista:** Rio Claro Campus and Proposal for Equipment to Treat Waste through the Composting Process. 72 p. 2009. Monograph (undergraduate) - Environmental Engineering Course, São Paulo State University (UNESP), Institute of Geosciences and Exact Sciences, Rio Claro, 2009.

SENAI/RS. NATIONAL INDUSTRIAL LEARNING SERVICE/RS. National Centre for Clean Technologies. **Implementation of Cleaner Production Programmes.** Porto Alegre: SENAI-RS/UNIDO/INEP, 2003.

SENAI./RS. NATIONAL INDUSTRIAL LEARNING SERVICE/RS. National Centre for Clean Technologies. **Environmental Issues and Cleaner Production.** Porto Alegre: SENAI-RS/UNIDO/INEP, 2008.

SOUZA, F.M. **Waste production control in a food and nutrition unit of a large hotel:** the importance of the nutritionist's role in the process. 2008. 19 f. Monograph (specialisation) - University of Brasília, Specialisation in Gastronomy and Health, Brasília, 2008.

SILVA FILHO, Antonio Romão A. **Basic manual for planning restaurants and industrial kitchens.** São Paulo: Varela, 1996.

SOUZA, M.F. et al. Characterisation of solid waste generated in a university restaurant. In: CIC, 18; Mostra Cientifica, 1, Pelotas, 2010. **Proceedings...,** Pelotas, 2010. Available at: <http://wp.ufpel.edu.br/rhima/files/2010/09/CE_00861.pdf>. Accessed on: 05 May 2016.

SILVA, Queli Viviana da. **Analysing the Application of P+L Tools in a Real Estate Paint Company.** 2010. 141 f. Dissertation (Master's) - University of Vale dos Sinos, Faculty of Civil Engineering, Postgraduate Programme in Civil Engineering, São Leopoldo, 2010.

SILVA, Patrícia Sardão da. **Application of the P+L and Leanand Green concepts in a concentrate dosing centre.** 2014. 118 f. Dissertation (Master's) - Universidade do Vale dos Sinos, Faculty of Civil Engineering, Postgraduate Programme in Civil Engineering, São Leopoldo, 2014.

SILVA, S.H. O surgimento dos restaurantes na cidade de São Paulo. **Revista Eletrónica de Turismo Cultural,** v. 2, n. 2, p. 05-26, 2008.

UNEP/UNIDO. **United Nations Environmental Programme/ United Nations Industrial Development Organisation** (2004). Available at: <http://www.unido.org/unido-united- nations-industrial-development-organisation.html/>. Accessed on: 05 March 2016.

TEIXEIRA, S.M. F. et al. **Administration Applied to Food and Nutrition Units.** São Paulo: Atheneu, 2006.

TEIXEIRA, Suzana Maria Ferreira Gomes et al. **Administração aplicada as unidades de alimentação e nutrição.** Rio de Janeiro: Atheneu, 1990.

THOMAS, J.M; CALLAN, S.J. **Environmental economics:** applications, policies and theory. São Paulo:

Cengace Leaming, 2007.

TINOCO, J.E.P.; KRAEMER, M.E.P. **Accounting and Environmental Management.** 3.ed. São Paulo: Atlas, 2011.

APPENDIX A - CONSENT AND COMMITMENT FORM FOR THE USE OF INSTITUTIONAL DATA

TERMO DE ANUÊNCIA E COMPROMISSO PARA UTILIZAÇÃO DE DADOS INSTITUCIONAIS

A ....GRSA localizada na Rua dos Andradas 1001, 7ª andar sala 701, Ed Gboex , Centro Histórico, Porto Alegre RS CNPJ 02905-110-0085/36, sob responsabilidade imediata de ....**Regina Bechelli** , **Gerente Regional - RS** compromete-se em ceder informações e dados para o projeto de pesquisa, intitulado:............**Aplicação dos Conceitos de Produção Mais Limpa em uma Unidade de Alimentação e Nutrição**.. .... com pesquisador responsável............**Ana Paula Bandeira de Oliveira** , colaboradora GRSA, para o **Pós Graduação Strictu Sensu – Mestrado em Avaliação de Impactos Ambientais**, Instituição/Setor: **Unilasalle Canoas.**, os quais também se comprometem a preservar a privacidade dos respectivos dados.

Os pesquisadores concordam, igualmente, que estas informações serão utilizadas única e exclusivamente para a execução do presente projeto, não podendo ser utilizadas para nenhum outro fim, sem a autorização individual e expressa das partes envolvidas.

As informações somente poderão ser divulgadas de forma anônima, garantindo o sigilo da empresa e informantes.

_Porto Alegre____________, __O7__ de __março_____ de 2016.

____
Pesquisador Responsável

____
Responsável pela Empresa
(Nome/Carimbo)

APPENDIX B - WEEKLY MENU TEMPLATE

	SECOND	TUESDAY	ROOM	FIFTH	FRIDAY
MEAT 1	GOULASH	GRILLED STEAK	BAITS WITH VACANCY	GRILLED STEAK	BAKED MEAT
MEAT 2	GRILLED CHICKEN	ROAST PORK SHOULDER	SASSAMI WITH WHITE SAUCE	NUGGETS IN TARTARE SAUCE	GRILLED CHICKEN
MEAT 3	MIXED BURGER	BISTECTION MUSTARD SAUCE	FISH GRILLED	SWEET AND SOUR SIRLOIN	ROSE SAUCE KIBBEH
GUARNIÇÃO 1	VEGETABLE SOUP	VEGETABLE SELECTION	SAUTE POTATO	BUTTERED COUVE	BERINGELA DORE
GUARNIÇÃO 2	RED SAUCE PASTA	SOFT POLENTA WITH SAUCE	RICE DUMPLING	STUFFED AIPIM	CARBONARA PASTA
SALAD 1	APPLE WITH YOGHURT	NECKLACE	TUNA WITH POTATOES	PASTA MAYONNAISE	FRUIT MISCELLANEOUS
SALAD 2	ALFACE	AGRICULTURE	CHICORY	COUVE CHINESE	MIXED GREEN
SALAD 3	OKRA	BROTO ALFAFA	BEETROOT WITH RICOTTA	RUCOLA WITH DRIED TOMATOES	WALDORF
SALAD 4	TOMATO	TOMATO AND ONION	TOMATO	TOMATO HALF MOON	CHERRY TOMATO
SALAD 5	COOKED CARROTS	CHUCHU	COOKED COURGETTE	RATATOULE	BEETROOT STICKS
DESSERT	PASSOQUINHA	CHOCOLATE PUDDING	BICOLOR CREAM	BISCUIT PIE	NEAPOLITAN MOUSSE
FRUIT	ORANGE	APPLE	BANANA WITH CINNAMON	MELANCIA	RODELA PINEAPPLE
GELATIN	STRAWBERRY JELLY	LEMON GELATIN	PINEAPPLE JELLY	CHERRY JELLY	GRAPE JELLY
JUICE	ABACAXI	UVA	MELAO	MANGA	GUARANA
DIET JUICE	UVA	MELAO	FISH	ORANGE	ACEROLA

APPENDIX C - SPREADSHEET OF ENVIRONMENTAL ASPECTS AND IMPACTS OF THE UAN UNDER STUDY

ACTIVITY	ENVIRONMENTAL ASPECTS	DETAIL OF ACTIVITY/ ASPECT	ENVIRONMENTAL IMPACT
receiving raw materials and inputs	solid waste disposal	paper, cardboard, plastic (secondary packaging), wood, light bulbs, styrofoam and organic waste (from fruit and vegetables)	Contamination/alteration of soil quality
storage of raw materials and inputs	electricity consumption	operation of freezers, refrigerators, refrigerated and frozen chambers, lighting, inspection of stored food	Depletion of non-renewable natural resources
administrative and support activities	electricity consumption	room lighting and operation of office equipment	Depletion of non-renewable natural resources
process (pre-preparing food, sanitising fruit and vegetables, preparing food, preparing milk formulas, enteral diets)	electricity consumption	operation of equipment and lighting	Depletion of non-renewable natural resources
process (pre-preparing food, sanitising fruit and vegetables, preparing food)	water consumption	washing, sanitising and cooking food	Depletion of non-renewable natural resources
process (pre-preparing food, sanitising fruit and vegetables, preparing food)	solid waste disposal	paper, cardboard, plastic, cans, glass (primary packaging), chemical product	Contamination/alteration of soil quality

		packaging, organic waste, packaging with sanitising product residues, glass/ceramic/porcelain from broken utensils, disposable material (disposable gloves, aprons, etc.) plastics)	
good manufacturing practices	reduced risk of physical, chemical and microbiological contamination	compliance with good practice criteria (time and temperature conditions, use of chemical products for sanitising fruit and vegetables and disinfecting equipment, utensils and the environment, and others, procedures for sanitising utensils, equipment and the environment)	Preserving the health of users and employees
distribution	solid waste disposal	clean leftovers, decorations, paper and plastic clean leftovers from events organised by contracted suppliers	Contamination/alteration of soil quality
sanitising equipment, utensils and the environment (administrative areas, storage, pre-preparation, preparation, distribution, return and service)	water consumption	general cleaning (dishwasher, dilution of products, various equipment and others), preparation of sanitising solutions	Depletion of non-renewable natural resources
sanitising equipment, utensils and the environment (administrative areas, storage, pre-preparation, preparation, distribution, return and service)	electricity consumption	dishwasher, combi oven, floor cleaning equipment (floor polisher) and lighting	Depletion of non-renewable natural resources (RNNR)
sanitising equipment, utensils and the environment (administrative areas, storage, pre-preparation, preparation, distribution, return and service)	solid waste disposal	waste from cleaning the grease trap, chemical product packaging, glass/ceramic/porcelain from broken utensils, water filters, cleaning materials, decorations from events (posters, banners, artificial flowers, etc.)	Contamination/alteration of soil quality
good manufacturing practices	reduced risk of physical, chemical and microbiological contamination	compliance with good practice criteria (time and temperature conditions, use of chemical products for sanitising fruit and vegetables and disinfecting equipment, utensils and the environment, and others, procedures for sanitising utensils, equipment and the environment)	Preserving the health of users and employees
rubbish collection, storage and disposal	solid waste disposal	temporary rubbish storage	Contamination/alteration of soil quality
disposal of finished preparations	water consumption	cooking, sanitising	Depletion of non-renewable natural resources
disposal of finished preparations	electricity consumption	operation of equipment and lighting	Depletion of non-renewable natural resources
disposal of finished preparations	solid waste disposal	Clean leftover food	Contamination/alteration of soil quality

APPENDIX D - FLOOR PLAN OF THE UAN UNDER STUDY

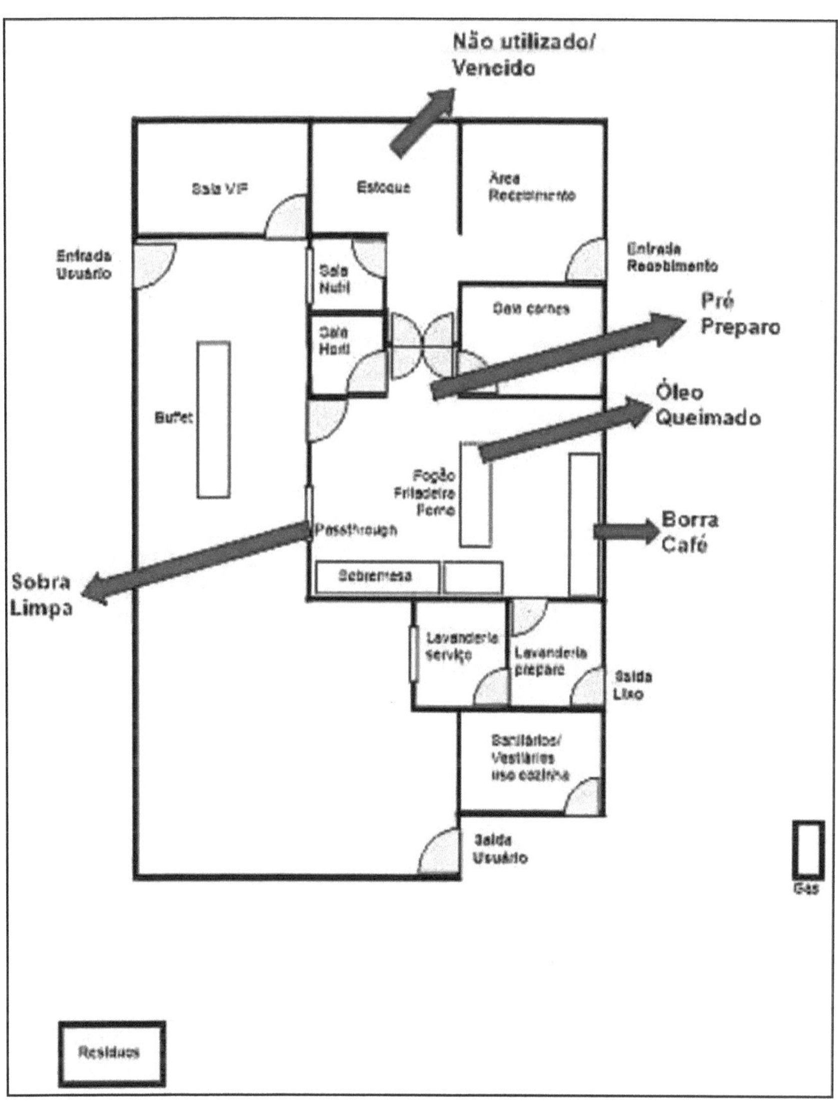

I want morebooks!

Buy your books fast and straightforward online - at one of world's fastest growing online book stores! Environmentally sound due to Print-on-Demand technologies.

Buy your books online at
www.morebooks.shop

Kaufen Sie Ihre Bücher schnell und unkompliziert online – auf einer der am schnellsten wachsenden Buchhandelsplattformen weltweit! Dank Print-On-Demand umwelt- und ressourcenschonend produziert.

Bücher schneller online kaufen
www.morebooks.shop

Printed by Books on Demand GmbH, Norderstedt / Germany